# ENGINEER BY DESIGN

*The story of my career from Apprentice to Chartered Engineer*

BY

PETER WILLIAM GEACH

*(Formerly Chartered Engineer and Member of the Royal Aeronautical Society)*

Author's Copyright - January 2019

Some years after retirement I resigned my Membership of the Royal Aeronautical Society because the 'Society' after always having traditionally allowed a reduced annual subscription rate for retired members; decided to abolish that category of membership. Retired members would in future have to pay the full fee; virtually doubling the previous long standing reduced fee for retired members. Working Members usually have their fees paid by their employer or if they pay themselves can claim against tax, which those in retirement can't.

Not being within reach of a dynamic Branch without a ferry crossing of the Solent, and practically out of reach of Hamilton Place for evening lectures, I was paying a lot of money for effectively the monthly Journal. I would have continued my Membership at the concessionary retired rate; but I couldn't justify the increased expense in retirement. Inflation would also see this escalate year on year; it was either resign now or likely be forced to in the future. I bit the bullet and resigned.

I thought the 'Society' were uncaring; after all, their reputation had been built, established and sustained to a large extent by the expertise of past members. Some accommodation I felt could have been made by the 'Society' if the will had been there; a sad event at the end of my career. As an apprentice, Chartered status had been the pinnacle of my professional aspirations, which I worked hard to achieve.

Something of which I was rather proud having left school at 15.

Clearly the 'Society' hadn't realised; or cared, that in Retirement, Professional fees were not tax deductable; as is the case for working members, who in many cases have their fees paid for them by their employer.

# TABLE OF CONTENTS

Foreword .......................................................................... 6

1. Introduction ................................................................. 9

2. APPRENTICESHIP @ Saunders Roe Limited, East Cowes .... 19

3. DESIGN DRAUGHTSMAN @ English Electric Aviation Limited, Warton. ............................................................ 34

4. DESIGN DRAUGHTSMAN @ F.G.Miles Engineering, Shoreham by Sea ............................................................ 49

5. WEIGHTS ENGINEER @ Dowty Rotol Limited, Cheltenham. .................................................................................. 57

6. SHIP DESIGN DRAUGHTSMAN @ MOD, Bath. ................. 62

7. PROJECT SUPPORT ENGINEER @ Bristol Siddeley Engines Limited, Patchway. ......................................................... 66

8. STRESS ENGINEER – POWERPLANT INSTALLATION @ Bristol Siddeley Engines, later Rolls Royce (1971) Ltd, Patchway. .................................................................................. 74

9. EARTHQUAKE ENGINEER @ New Zealand Electricity, Wellington, NZ. ............................................................. 85

10. Working In The UK ..................................................... 106

11. SENIOR STRESS ENGINEER @ British Hovercraft Corporation, East Cowes. ............................................. 108

12. STRESS TEAM LEADER @ Straughan & Henshaw, Bristol ........................................................................................... 110

13. PRINCIPAL MECHANICAL ENGINEER @ Marconi Space Systems, Limited, Portsmouth. ..................................................... 117

14. SENIOR STRUCTURAL DESIGN ENGINEER @ Vosper Thornycroft (UK) Limited, Woolston. .......................................... 134

15. Making Your Own Luck. ....................................................... 149

16. The Best & Worst Firms ........................................................ 153

17. Reflections............................................................................. 154

# FOREWORD

This has been written as much as anything else to provide a social history of Engineering and the opportunities it offered in a lost age to young people prepared to study, and seek out the opportunities open to them during the 1950's until the 1970's.

At this time, Moneytryism, Outsourcing and Technology Transfer to the EU took precedence. As a result, Industry contracted, and these opportunities withered and disappeared, as did many of the firms offering them.

During these brief 30 years there was outward and upward mobility for ordinary people with intelligence, ambition and prepared to study; having self- reliance making up for external influence.

The typical route for a craft apprentice wanting to progress into the Drawing Office was obtain the Ordinary National Certificate (ONC) in Engineering during the Day Release Scheme at Technical College. A Higher National Certificate (HNC); the next grade up would guarantee that, and offer further progression into the Technical Offices; the Stress Office for instance. At the end of a 5 year apprenticeship the HNC was within reach and often achieved.

I took this route, but went on a unique journey in the process, moving employer and upgrading my job and gaining experience along the way. I lived and worked in New Zealand for 5 years before deciding the UK was best after all; but it was a wrench.

This is my story within that caring opportunistic time and what came after. I have tried to show how an ambitious person could significantly improve their job and prospects.

Although those days of opportunity are long gone, however with initiative and belief in themselves, and studying for where they want to be, should help in finding a way forward. It's far from easy these days.

Along the way I have included a potted history of the firms and organisations I worked for. Most of these companies have been absorbed and names lost; which happens when work becomes scarce.

Our hard-won Aerospace Industry and expertise will never recover, the Politicians have seen to that. The question is was it done intentionally to keep people down, or through folly and ignorance? Probably a bit of both. What do you think?

I've just recorded what happened; it's for you to decide the rights and wrongs!

Lack of investment and cutting back on apprenticeships has been the major cause of our decline as an Industrial Nation. Outmoded, out of date factory layouts and machinery have been the route cause.

No one can compete say, if metaphorically speaking you compare the relative productivity of pumping up car tyres with a foot pump, with that of somebody using an air compressor! The technological gap produces the same result! Old factories, outdated layouts, procedures and equipment just can't compete with new purpose built 'state of the art' factories.

Today Governments give lip service to Apprenticeships. They describe semi- skilled on the job training schemes as an Apprenticeship! Either Politicians are ignorant of the real world, or liars and conmen. In either case they shouldn't be in Parliament looking after the Nations well being.

We've seen the Government folly of pressurising Vosper Thornycroft to build a 'state of art' ship construction facility at Portsmouth Dockyard for the Type 45 Destroyer and Carrier program, for the company to be later forced by the MOD to merge with and the ultimate takeover by BAE Systems, based in Glasgow, where work was transferred and the new Ship Building facility in Portsmouth closed. Shades of TSR2!

Vosper Thornycroft, when based at Woolston on the River Itchen near Southampton under their Managing Director Martin Jay, invested heavily in chasing and obtaining world sales for Naval Corvettes and High Speed Patrol Craft. This was done on the back of their Single Role GRP Mine Hunter ordered for the RN and Saudi Navy.

Mr Jay also expanded the company into Nuclear Engineering, Royal Navy warship maintenance and repair at Portsmouth Dockyard and developed Fleet Training; it became a multi-role company. I was always surprised he didn't receive a knighthood for his enterprise and vision! But hey, that's reserved for the luvvies!

And now of course the Governments mess up of BREXIT! No factual debate; much misinformation. Neither the Government or Labour opposition countered that spin. The referendum result was 52% for; 48% against. The unsubstantiated claim that on leaving the EU, the NHS would have £350,000,000 extra funding per week probably swung the vote! Someone is going to make a lot of money; many will lose out heavily including Industry, Banking the pension funds, pensioners and population as a whole, except the favoured few!

Surely the job of Parliament and MP's is to ensure there is investment and real opportunities for skilled, well paid work to remain alive for everyone; not just the influential big boys! MP's should by statute have to come forward from good high level jobs where they have been successful. That would ensure only the best candidates could stand.

I've used Wikepedia on the web to research the firms and P D Stemps book Kites, Birds & Stuff – MILES Aircraft. Ref ID: 8200186 www.lulu.com

# 1. INTRODUCTION

Engineer is a much over used term in the UK. Anyone working in an engineering related occupation; to the general public anyway; is an Engineer; be that servicing the washing machine, car or repairing burst water mains. Media please note!

That makes as much sense as calling anyone working in the health service a Doctor! As in the health service, there are also a number of grades of job in Engineering; from the craftsman to managing director. Relatively few are categorised as Engineers; many are craftsmen and technicians. To use the designated title 'Engineer'; that person has to be Chartered.

This means being educated to degree level in a specialisation plus recognition of appropriate training, skills, experience and a demonstrated level of responsibility in that specialisation; overseen by a Chartered Engineer, before the applicant can be assessed by a review board for Chartered Engineer status and membership of an 'Institution'.

The process takes some years, even for those with a first class degree. So the term 'Engineer' is not something that should be banded about loosely by anyone working in engineering, the media or general public.

Chartered Engineer status is not necessarily only achieved by those taking the degree or HNC/D (Higher National Certificate/ Diploma) route.

A Craft or Technician Apprenticeship with appropriate B.TECH and Higher B.TECH (ONC and HNC in old money) qualifications from day release at Technical College, with appropriate additional endorsements meeting the technical requirements of the appropriate Engineering Institution is another route. A further option is self study and sitting that Institutions entrance examination.

But in each case the appropriate experience and level of responsibility in the job has to have been demonstrated as outlined above.

An HND was a coveted qualification, awarded by Polytechnics to students having completed a 'Sandwich Course'; half a year in Industry with a sponsoring employer and half at Polytechnic. The HND was a degree in effect by another name, awarded by a Polytechnic rather than a University. Employers regarded the Polytechnic HND qualification highly, because the graduate with an HND had obtained good Industrial experience along the way; was a known quantity and initially more useful than the University graduate.

Employers able to give the appropriate training, experience and professional supervision is a requirement to obtaining Chartered status on top of the required academic qualifications for membership. Appropriate employment is something not all aspiring craft and technician apprentices may find easy to obtain; if for instance the firm is relatively small and not able to offer the opportunities of training and progression through the company. In which case after completing apprenticeship you have to move on to a company that will train you up to chartered status.

It's worth briefly running through some of the various job categories to try and clear up popular misconceptions. I have also added a potted history of my early career to illustrate how the system worked in my day and how I fitted into the scheme of things.

**Craft Apprentice:**

Skilled craftsman with craft qualifications obtained at Technical College Day Release and evening class. This could eventually lead to: Charge-hand, Foreman, QA, Inspection, etc. or into the Production Offices as Planner, Jig and Tool draughtsman etc. But essentially Production orientated.

Craft apprentice's may be able to transfer into the Technician Apprentice category if they obtain good Technical College results;

say B.Tech (ONC), and the firm has a progressive policy and the openings are there. This needs hard work, belief in yourself and some luck. Technician Apprentice status opens the door to the Drawing Office.

**Technician Apprentice:**

Usually technician apprenticeship places are reserved for those school leavers with good academic results. But there used to be a 'safety net' in that good results at 'Tech'. can earn Technician Apprentice status. It's something the individual apprentice will have to take up with the Training Department, and pitch his case for technician status.

Have realistic ambitions in the area of work and job you want to end up working in at the end of training. It's surprising what a sensible talk with a sympathetic training manager can achieve. It could be along the lines you get this qualification at Technical College, and I'll try and get you into the Drawing Office and see how it goes! Can't ask fairer than that! But if he's unreceptive, at least confirm the qualifications needed for your aspiration and work steadily towards it.

Technician Apprentices study BTECH (ONC) and Higher BTECH (HNC) on Technical College Day Release and evening class. These qualifications can take the BTECH (ONC) qualified technician into the Drawing Office or Planning say; the Higher BTECH (HNC) technician into the Technical Offices (Structural Analysis say), Project Management and so on.

Plenty of scope and opportunity then for apprentices with the Higher BTECH. Which could over time with endorsement qualifications, experience and responsibility lead to Chartered Engineer Status.

**Graduate Apprentice:**

The Graduate Apprentice came with an HND or degree and went through a graduate apprenticeship training period; probably 3 years; with time in the workshops, production and technical departments. The graduate's attitude and performance was monitored and towards the end of his time offered a placement of his choice if available or chance of transfer later.

Training and supervision will continue with a view to gaining Chartered Engineer status; important not only for the graduate, but company too because their technical competence to undertake a complex project, will in part, be judged by the number of Chartered Engineers; professionally qualified people; on its payroll. And because this status is important to the company, the firm often pays professional fees.

The Graduates training period throughout the company is very important both to the graduate and the company. The graduate learns how the company functions, and how things work; gets to know the really good people in the various departments whose help he can trust. He'll also know who it's best to keep clear of!

### **Chartered Engineer:**

This stage is reached in the late 20's at the earliest. The appropriate skill levels and experience won't have been reached till then.

Is the HND or degree qualified Chartered Engineer, better than one who has come through the Technician Apprentice route with Higher BTECH and endorsements?

In general the young chartered engineer with the industrial background will probably be the better all-round package in the technical support role. Outstanding apprentices can go onto high levels in the company.

The Polytechnic and University qualified chartered engineer would be better academically. The best will be fast tracked.

Experience and speciality over the years can even out. However; the academic edge is likely to stay with those who've had extended studying opportunities at polytechnic or university. It really depends on the job and the individual. If you are not sure about something there are books, colleagues and the internet.

Good, sound experience, knowing your own and Code limitations is more important than personal brilliance. Overconfidence can lead to disaster.

**<u>Getting Started:</u>**

It's often difficult to know how to get started; climb the greasy pole if you like; especially if you aren't clear on what you would like to do, or academically shy! Schooling; so unfair, but there it is. Fortunately, the Technical College provides a 'safety net' for those willing to get their heads down!

I hope the preceding explanation of the various categories and training routes has answered most questions. These routes were available in my day; some may not be 'on the menu' now and take special perseverance.

Do as well as you can at school; think of topping up qualifications at local Technical College. But first, find out what you need for University and for a Technician Apprenticeship, and if apprenticeships are likely to be coming up locally.

Find out about a company's training policy; is it progressive; could you transfer from craft to technician apprentice at a certain academic level at Technical College?

Could you move into design say when you have obtained a certain academic qualification? If this has been discussed at interview, important to get your understanding of this confirmed in writing from the firm! If this possibility is not included in the 'offer letter'; write saying what you understood was promised and would they please confirm.

University brings in its wake huge expenses to pay back; so think hard before you commit. Loan repayment will impact on mortgage and setting up a home in the future. Although Martin Lewis (MONEY Program), says if 'Uni.' is right for you, take the opportunity; think of the repayment as a tax, after 30 years the loan is void, you pay back according to salary, so may not over the 30year loan period pay everything off; if not paid back in 30 years, nothing more to pay; so a good deal. But do check it out for yourself.

There is no guarantee these days a degree will find you a good job. So be very selective in what you study; the more academically based the better; and be sure it gives you options after qualification.

For instance, a Structural Engineering Degree would find you well paid employment in Civil Engineering, Aerospace, Shipbuilding, etc. anywhere structural integrity matters; and it's relatively easy to cross over from industry to industry.

Whereas an Architect say, is pretty much limited to the Building Industry and Planning Office, and generally isn't much use as a structural designer in say Aerospace without serious re-training. If the Building Industry is depressed and the Architect is made redundant, future work prospects are severely limited, as would be his salary in a new job.

If university bound look around for a firm that may be interested in sponsoring you in part, offer you work during holidays and a graduate traineeship. That firm may not necessarily be close to home! These companies need to be searched out; they won't come looking for you! A career in the Armed Forces should not be overlooked.

With about half of school leavers going to University these days; there is no way Industry and Commerce, NHS, Local Government etc. can absorb this number of graduates year on year; nor give them the career path and salary progression they seek! Just finding a job is difficult.

Where are the jobs these days? Look around!

In the 1960's and 1970's Industry and Commerce was still expanding; there were fewer places at University, but when graduating there were good career jobs to go to. The Grant System ensured students didn't build up huge debt. Totally different today; the jobs just aren't there anymore. Never would have been, even then, for the number of students graduating each year nowadays!

So, don't overlook the Technician Apprenticeship route if it will get you where you want, with no big fees to pay back! And after training, most likely, a career with that company. Having invested in your training the company will want to retain you. You are a known quantity and know the ways of the firm; win, win I'd say!

Always go for the widest options that give you most choice when finalising your study choices at school. Don't get specialised too soon, before you really know what will be available to you. The trick is to gain wide experience early with good firms and specialise later when you know what you want.

Believe in yourself and good luck. If you read through my story you will see how to position yourself and make your own good luck; not straight away, but in time.

It's worth repeating the story of the man with fixed ideas of his career path. He was walking down a long corridor with open doors of opportunity. With only eyes on the end door, he walked past. When he got there it was locked. He turns to walk back and all the other doors, open on his way up were now also closed!

The moral of the story is not to tie yourself rigidly into any one idea too soon; it may never happen; be sensible and flexible. Always leave yourself options where you can. There is usually more than one route to achieving your ambition.

Getting started in a company with a progressive training structure is what you should look for above all else. Keep your eyes open at the end of training, when you should have found out what you are really interested in. If not with that firm, you know the job to look out for; and go looking!

## **Me!**

I left school at 15 having taken no exams; a type of education many working class kids endured. I didn't really know what to do; I'd toyed with the idea of applying to join the Royal Navy as an Artificer. My Dad who had been to sea advised against it, didn't think I'd like the life. Then I thought of being an Apprentice

Boatbuilder; but the best place of all for a craft apprenticeship was Saunders Roe who were building and testing the Princess flying boats; I saw the launch and days later first take off as the Queen Mary was sailing up the Solent to Southampton.

I won a Saunders Roe apprenticeship under the charismatic Apprentice Superintendent of Training, V T Stevenson, known to everyone as VT. The Apprentice School was housed at Whippingham in premises used previously by the Royal Navy Osborne Cadets for sleeping quarters. The school used 4 large dormitory buildings; one refitted as the Technical School with drawing boards, spare space was taken up by Administration Offices; one was converted as a Fitting and Machine Shop. The remaining two Dormitories were kept in use to accommodate Apprentices from the Mainland. VT lived in the Commandants old premises. Later the Rev. Ron Scruby joined the company as Chaplain to the apprentices; his office was situated in VT's premises where VT also had his office. Ron Scruby had lost a leg on D-Day; took up rowing at University, had two sea rowing gigs built and introduced the sport to the apprentices. A large Gymnasium and Library was built on the site. Judo and other sports were taught there. Ron Scruby and his wife lived in a flat on Cowes sea front.

The apprentices also had access to an Engineering Graduate attached to VT's staff who acted as tutor one evening a week in one of the offices if any of us needed help with Technical College work. That facility was open to all apprentices, not just those living on site in the dormitories. I used to cross the River Medina from Cowes and cycle a couple of miles from time to time to take advantage of this help.

VT was inspirational; made us all feel 6 foot tall; especially me who was mixing with Grammar and Public School Boys; University in those days was for the few; the very brightest and those who could pay.

VT opened up the field of Engineering to us, rather like opening Pandoras Box. He made it clear how we could progress either through the craft City & Guilds exams. Or if we wanted to progress into the Technical Offices described the Ordinary National and Higher National Certificate academic route to eventual Chartered Engineer status; wow!

This is just a taster of the company I trained with; I'll expand on this later; probably repeating myself here and there to make it cohesive.

For what he did over a number of years, VT deserved public recognition. If he got any it was only a minor 'gong'. Knighthoods are reserved for 'luvvies' who often, I think, get overly full recognition for their talent and very well rewarded financially.

I think some form of public recognition is also due to outstanding college lecturers, who by their skills turned many youngsters lives around. I could go on.

The Isle of Wight Technical College provided the academic stimulus through such lecturers as Jayden David Phillips (JD), the Sydenham brothers St. Barb and Mike. Marcus Langley specialised in Aeronautics; the wind tunnel ducting ran through the Ladies Toilets; no discrimination in those days!

There were other lecturers whose names I forget, but they were all distinguished by knowing their subject, making it interesting, treating everyone with respect, and boy could they teach! Never had anything like it before at school. Suddenly I found I could do this stuff; I had to work hard playing catch up because of my schooling, but I got good results usually in the top quartile.

I started in the Fitting Shop learning how to cut out a 5 inch square from rusty steel plate and file it up square within a few thousandths (thou) of an inch. The surfaces also had to be parallel and flat. Sticky plastered fingers abounded on those soft little hands!

For a change we made complicated flanged brackets from aluminium plate. We had to calculate bend allowances, mark everything out and cut out the developed planform; make special bending tools out of 'Jabrock' (a hard plastic type material) and bend up the item. It had to be made within close tolerances, say 15 thou! If not try again until it was right.

We made hand tools and learned to use the Lathes and Shaping Machines.

We were given a break from this with a spell in the Offices as part of pre-apprentice training. I went to the Drawing Office at Osbourne as Office Boy. This meant running errands for the draughtsman, getting prints etc. I was sat next to the Section Leader Mr Kennedy. He saw I was keen and got me to do a small drawing; a spacer; glorified washer! But I was chuffed.

I regularly rode my bicycle between the DO and Technical School; about half a mile along the road to see 'Poodge' Saunders the Technical Instructor, for extra tuition.

He would set me Geometry exercises which I would do at my desk in the DO and when completed bring them back for him to check! I was keen. I decided it was the DO for me, my sights were set on the Ordinary National Certificate (ONC); the first step to becoming a design draughtsman.

# 2. APPRENTICESHIP @ SAUNDERS ROE LIMITED, EAST COWES

The best option open to me as a school leaver was an Apprenticeship at Saunders – Roe Ltd. East Cowes, across the River Medina from where I lived in Cowes.

The firm designed and built flying boats; their latest was the Princess flying boat, a design started in 1943; a 747 sized aircraft way back in 1952 when it first flew, a 200 passenger 350mph transatlantic flying boat. The Princess was undergoing flight trials when I joined the company. They also had a projected all jet flying boat; the Duchess, powered by 6 de Haviland Ghost engines; it was a swept wing design and looked a little like the Martin Sea Master jet flying boat. The Saunders Roe Duchess project wasn't proceeded with after BOAC cancelled the Princess and stopped flying boat operations.

The company designed and built Space Rockets, and under contract, built Supermarine Swift and Scimitar tail-planes as well as the Vickers Armstrong Valiant V bomber pressure cabins; a busy vibrant company.

Design was well advanced on the Saunders Roe mixed rocket/jet propelled interceptors. A prototype was in final stages of construction.

Saunders Roe also had a Helicopter Division, based in Eastleigh, building the Skeeter Army Observation helicopter.

In 1928 Alliot Verdon Roe bought a controlling interest in S.E. Saunders Limited. A year later the company name was changed to Saunders Roe Limited.

Sam Saunders had a fast launch boatbuilding business on the Thames at Streatley and Goring. He sold the business and moved to East Cowes on the River Medina in 1908, calling his new company

S.E.Saunders Limited. A year later he started an aviation department.

Whilst his main business was designing and building high speed launches, and record breaking power boats, the company started designing and building their own aircraft, principally float planes and flying boats.

During the First World War, the company built flying boat hulls and fighter aircraft for other manufacturers.

Saunders Roe built Miss England II for Sir Henry Segrave with which he won the absolute speed record on water. Sir Henry Segrave was later killed in 1930 when at speed on Lake Windermere the boat struck a submerged object, overturned and sank. The boat was raised, repaired and went on to raise the record further in 1931.

In 1937 Saunders Roe built Sir Malcolm Cambells, Bluebird K3, a boat which took the water speed record that year. A later boat, Bluebird K4 was built by Vospers Limited at Porchester (near Portsmouth), increased the water speed record in 1939. I mention this because there is sometimes confusion over who built the later boat.

In the early 1940's, the company designed a twin jet flying boat fighter. It was intended for the Pacific War. Three were built. It first flew in 1947, too late for the war. One ran into a log on landing on the Solent, and sank. One survives in the Southampton Aviation Museum.

In 1952, rearmament was in full swing. It was the Cold War period; the shops were busy building Valiant bomber pressure cabins, Swift and Scimitar fighter tail-planes. Saunders Roe were also involved in designing, building and the testing of Space Vehicles; the Black Knight Launch Vehicle, eventually with the Black Arrow atop for launching small satellites into space orbit. The Space Vehicles were manufactured in Folly Works by the River Medina, test fired at The Needles Test Site, before being sent to Woomera, Australia, for launch trials.

The company was also busy designing and building prototypes of a mixed jet and rocket interceptor for the RAF; the SR 53 development aircraft and it's successor, the larger more powerful, production variant the SR 177. A very busy, and expanding company in those days.

Saunders Roe were at the spearhead of Aerospace technology of the time. They had about 5000 employees. A remarkable achievement for a relatively small firm. No doubt that was the place to be apprenticed, if you could get in.

The Apprentice Training School was set up by V.T Stevenson (VT) during the second world war to train apprentices and unskilled people in the skills of aircraft production.

After VT retired in the late 1970's I suppose; Roy Jones took over. He maintained the high standards and developed training further for the Youth Training Scheme (YTS).

After final retirement VT emigrated to live close to relatives in South Island, New Zealand where he died; fortunately, before the Training School was closed; that would surely have broken his heart; he'd put so much into it, it was his life's work.

A few years after Roy Jones retirement in the early 1990's the company; very short- sighted; I thought; shut the Training School down, losing the Course Curriculum, the spirit de corps it engendered in all the apprentices and not least the collective skills of the Instructors, some of the Company's best craftsmen. How short sighted can a Board of Directors be? It has been said the company were paranoid about preservation orders being placed on the Naval Cadets dormitories and resulting costs, so closed the Training School, demolished and redeveloped the site before that happened. I hope there was no truth in that rumour.

The company suffers today from that closure, in that apprentices coming through no longer have the same aptitude and skills set and spirit de cor.

A story of Accountants knowing the cost of everything, but the worth of nothing!

I sat the entrance exam; passed that and the subsequent interview. Saunders – Roe offered probably the finest apprenticeship training in the country, certainly in the South of England. I doubt it was bettered anywhere.

V T Stevenson, the Apprentice Superintendant of Training; I felt; was never officially recognised for the Training Scheme he set up and ran with such flair.

Apprentices came from the mainland and boarded at the Training School which was situated at Whippingham and used the old Osborne Navy Cadets dormitories. One was used as administrative offices and Technical Lecture Room set out as a drawing office, another housed the Training Machine and Fitting Shops. The remaining two; were used as Dormitory's for mainland Apprentices.

The company also had a Test Tank, Wind Tunnel, Test House, R&D offices and buildings on site, where flight systems and structural testing took place. Wooden mock-ups of future aircraft projects were built in a large shed on site.

As a schoolboy I remember the next door neighbour taking me with his children to see the Princess flying boat mock up and seeing the Princess being built in the Columbine workshop by the River Medina; even managed a quick walk around inside the fuselage! An inspirational outing, and one that got me interested in an apprenticeship with the company.

VT ran the apprentice scheme rather on the lines a 'House Master' may have done at a public school. He certainly instilled a 'spirit de corpe' amongst the apprentices. We were all made to feel rather special; this made us in return put our best foot forward and 'raise our game'.

Starting work at 7-45 each morning, with an hour lunch break and then on until 5-30p.m. was hard after genteel school hours. Morning Tea Break was 9-30, and that seemed an age after clocking on. After a day in the Lecture Room being told what was expected of us and the opportunities if we worked and got the exam results we needed; life began in the Training Fitting Shop.

The first job I remember was cutting up a piece of rusty 5/8 inch thick steel plate into about a 5 inch square. This was done by marking out and drilling a number of closely spaced holes along the sides, then cutting through the remaining metal with a hammer and cold chisel! The next task was to use a heavy cut 'bastard' file to roughly square off the edges and remove rust from the other surfaces. Then the fun began; each surface had to be filed flat, square and parallel to each other. Square and flatness was checked with a square; dimensions measured with a micrometer and vernier gauge for the larger dimensions. It was all done by filing using various grades of file. When putting on the final finish, chalk was sometimes rubbed on the file. Accuracy; an astonishing plus or minus a few thou! That is what the Instructors expected and that is what you achieved. You didn't do anything else until it was right! Poor sore fingers draped in sticking plaster! There weren't any girl craft apprentices in those days! That was before Women's Lib.

There were other exercises, making tools for yourself to drawing, using the same hand process, drilling and tapping, turning up bolts on a lathe, using a shaping machine.

Having learnt to work accurately within close tolerances in steel, we began plate bending exercises in aluminium alloy sheet. Bending allowances had to be calculated, based on the quarter circumference of a circle using the mid bend radius of the plate (inside corner radius plus half plate thickness). Special steel bend bars were used having the rounded edges of appropriate bend radius for specific thickness of plate, say 2G, where G was the plate thickness. The bending exercises were highly accurate and skilled work, the finished work had to be to plus or minus 15 thou! It could be done, but some exercises had to be repeated to achieve this accuracy.

We had two spells in the fitting shop. After 3 months we were put in the 'Offices' as office boys to see how the firm worked and the opportunities that may be open to us if we did well at Technical College. I was sent as office boy to the Aircraft Drawing Office along the road at Osborne for my stint.

I knew immediately I wanted to be a Draughtsman. The admission ticket was an Ordinary National Certificate in Mechanical Engineering; that was the goal I set myself from then on.

It was then back to the Training School fitting and machine shop to complete our skills training before going out to the shops and work proper.

The Isle of Wight Technical College was staffed with enthusiastic lecturers who knew their subject and how to teach.

I'd left school feeling academically unaccomplished, but at 'Tech' everything changed. The teaching was of a standard I'd never had before. The subjects were taught in such a clear way that you developed confidence and worked not only for yourself, but also because you didn't want to let the lecturers down at exam time! What a difference from school. For me it was the whole package; Saunders- Roe combined with an excellent local Technical College that made the difference; but I had to work hard 'playing catch up' with grammar school boys! It wasn't easy, but definitely worth it.

I began in an Introductory Year at Tech. Depending how you did in the course work and exams, in the following year you either went onto the craft based City & Guilds course, or if you did well in the preliminary year, onto the more academic Ordinary and then Higher National Certificate courses, with entrance to the Drawing, Stress and other Technical and Production Departments depending on your specialisation.

So, I got my head down and worked.

In the first year of Ordinary National (S1), I was stumbling a bit over maths. I put an advertisement in the local paper for tuition, and went once a week to Peter Baxter; a school teacher until I got up to speed. He said he'd met Jayden Phillips (my tech. college lecturer) and his brother at Winchester Teacher Training College. I paid for tuition out of my apprentice's wage.

I mention this to show that if you are determined enough you will always find a way. Have an aim, believe, and invest in yourself.

After the Apprentice Training School, I spent time in the Press Shop, setting up and using Fly Presses. Then it was the Test House at Whippingham where I made test fixtures, components and helped set up the Elephant Test Rigg where the SR53 tail-plane was to be tested to destruction; an interesting experience. I also observed load/deflection tests on the P531 (Wasp) Helicopter nose undercarriage.

Pressure testing was being carried out on short mini fuselage like pressure vessel structures as part of a fatigue test program. This entailed perching on a platform above a water filled tank, changing failed test pressure vessels for new. I raised a laugh by in all seriousness asking! 'if I fell in, did I go home to change in my time or the firms'! These fatigue tests were being carried out to further knowledge of fatigue; about the time of the Comet fuselage pressurisation fatigue failures.

My next move was to the Hydrodynamics Test Tank Fitting Shop, where I assisted with testing a model hydrofoil and a new design of self- righting lifeboat for the RNLI. In the test tank.

Looking back, I think the part I played in the Hydrofoil Testing was a little dodgy! My job was to wear long thigh Wellington Boots in the Test Tank with an electric cable on a long pole connected to an electric motor in the model hydrofoil, the other end to a bank of car batteries. I had to turn with the powered model, running in a circle, ensuring that the cable stayed connected!! What would have happened to me if the live cable went in the water doesn't bear thinking about. Health and Safety; never heard of them in those days!!

The inertia of the model Lifeboat was measured by the Pendulum method. It wasn't explained to me how the inertia was measured and calculated, or what effect the model inertia had on the tests. I suppose I was just a useful pair of hands; an Apprentice who wouldn't understand. You got that attitude sometimes from degree people in those days, an exclusivity from normal mortals who hadn't had a silver spoon in their mouth upbringing! That's how it was if you didn't have much status; handn't perhaps a Grammar School background; but not from VT, everyone was made to feel good about themselves; a great man.

Also, on site was the Ditching Tank. Here models of the V-bombers; fitted with accelerometers were launched by catapult to measure accelerations when they ditched in the tank. Many years later it was said a model of Concorde was tested, but the accelerations on ditching due to it's high angle of attack at low speed would snap the fuselage!

An idea to get over this weakness was to fit a water ski to the nose undercarriage to reduce accelerations on ditching; I think it was shown to work, but never proceeded with; probably impractical to incorporate at that late stage in the Concorde design.

Then it was Folly Works, an idyllic setting on the banks of the Medina River by the Folly Inn. It meant a very early start to catch the works bus at East Cowes. There, I made mock up parts for the SR 177; successor to the SR 53; which was a development aircraft, although originally, I think it was a prototype for a production interceptor.

The SR 177 was an improved development of that concept. At Folly Works, I also worked on Scimitar tail-plane assembly and on the Black Knight space rocket structure.

After I completed ONC, I was put in the Production Development Department where the Chemical Etching Process for aluminium skins was being developed. Not really what I wanted, but a step in the right direction. I'd have to see VT and make my case for the Aircraft DO sometime.

I'd started rowing in the company's gig, and competing in local regattas; was making good friends amongst the other apprentices and everything seemed set fair. I'd passed my driving test, bought an old Wolsey Hornet car, which gave me a summer's fun before it gave up the ghost. I then bought a small 197cc two-stroke Francis Barnett motorcycle which got me about.

The first year (A1) of HNC was going well and looked forward to progressing onto A2 the following year; then I'd certainly get into the Drawing Office.

But things suddenly changed; the 1958 Defence Review! In this review all manned bomber and interceptor projects apart from the Lighting fighter being developed by English Electric Aviation at Warton (Lancashire) were going to be cancelled. The Lightning was to be the last manned Fighter for the RAF. Combat aircraft would in future be replaced by missiles; that was the mantra of the time!

Up to that announcement, Saunders-Roe were about to take over Westland Helicopters. After the SR 177 cancellation, Westland Helicopters took over what was left of Saunders-Roe; which eventually became the The British Hovercraft Corporation developing hovercraft using the principal invented by Christopher Cockeral with Saunders-Roe.

Prior to the 1958 Defence Review; the SR 177 interceptor fighter was being developed for the Fleet Air Arm. The Luftwaffe showed interest. If they bought it, the development would continue, otherwise it would be cancelled.

The Luftwaffe; I guess under strong political pressure; chose the ill-famed American Lockheed 104 Starfighter instead; a project not flown by the US forces, but was flown in large numbers by European Air Forces, so I suppose became a standard type. And so the SR 177 which was at an advanced stage of construction in prototype form was cancelled.

It was my 4$^{th}$ year of apprenticeship. I needed a further 6 months of employment on top of my apprenticeship to see me through to the final year HNC exams. because my first year at Technical College had been spent in a preparatory class, for the ONC course. Normally with SR177 project in place I could have obtained deferment from National Service to complete HNC. But without SR177 I faced National Service and most probably unemployment on completion of my apprenticeship. The bottom dropped out of my world! I thought if I do National Service without completing my studies, would I ever be able to pick college work back up again. As soon as my apprenticeship was completed at the end of January I would be called up and miss the June HNC exam.

Usually; if you worked in the Defence Industry you could rely on at least two years deferment after apprenticeship to complete exams to meet entrance requirements for one of the Engineering Institutions; in my case The Royal Aeronautical Society. Then having completed National Service, and back at work again, with experience and responsibility, eventually achieve Chartered Engineer status.

Big changes were afoot for everyone. Westland took over Saunders Roe. I don't think the firm in all its guises since has ever recovered the spirit de corps of Saunders Roe!

VT Stevenson; the Superintendant of Training; wrote to all apprentices in their third and fourth year of apprenticeship with the proposition they truncated their apprenticeship's and completed National Service early, returning afterwards to complete their apprenticeship after national service.

I went to see VT with my father and explained my situation regarding the HNC exams. He could offer no prospect of even limited deferment until the HNC exam. Indeed, in the current situation the company was in, it would be unlikely the company would be able to employ me after my apprenticeship was completed. I privately wondered if it would even exist in a few years time!

But the firm survived through various re-incarnations. There were lean times but it survived.

British Hovercraft Corporation (BHC) was formed in 1966 with the hovercraft interests of Westland and Vickers Supermarine. BHC was based at East Cowes in the old Saunders Roe premises; where the main business was the design and manufacture of hovercraft. BHC remained part of the Westland Group.

Because of financial difficulties, the Westland Group including BHC, became Westland Aerospace in 1985.

GKN bought the concern in 1994, becoming GKN Westland. In 2018 the company was taken over by a business development company, hoping to sell off certain assets and make money. Too early to see the result of that strategy yet.

But getting back to my story; being very interested in aircraft, I often bought the Flight Magazine on a Friday and avidly read through it. I had noticed for some time that English Electric Aviation at Warton, Lancashire were advertising for suitable people to join their Trainee Draughtsman Scheme for their Aircraft Design Drawing Office at Warton, to work on the Lightning Fighter.

After having seen 'VT', and with no chance of deferment to take my exams, I wrote off to English Electric Aviation replying to their advertisement and enquired whether I could transfer my Apprenticeship to them and also join their Trainee Draughtsman's course. As a 'second string', I decided to write to Shell on the Isle of Dogs to enquire if I could transfer my Apprenticeship to them. That's called initiative and making your own 'good luck'.

I hadn't discussed this with 'VT' first. As I saw it, it was up to me, no help from anyone else!

A couple of weeks later I got a phone call from VT: Hello my boy; he called everyone 'my boy'; he said I have letters here from English Electric and Shell about the possible transfer of your apprenticeship. Come up and see me this afternoon and we'll talk about it.

When I saw him, he said I can arrange this for you; which firm do you want me to approach. I said English Electric at Warton please; that would get me into the Drawing Office and give me the chance to complete my studies. I felt if these were interrupted by National Service, I would find it very hard to get back into the studying habit. I'd worked too hard to get what I had achieved. And in any case after National Service would the company still be there? I could certainly kiss goodbye to the drawing office. VT said he would arrange an interview and wished me good luck.

English Electric wrote inviting me for interview at Warton. I travelled up by train and stayed overnight at The Midland Hotel, just across the road from Preston Railway Station. A car was sent for me after breakfast to take me to Warton for interview with Mr Daly, the Training Manager.

I was offered a place on the course with also the transfer of my apprenticeship from Saunders-Roe Ltd to English Electric Aviation Ltd. Mr Daly said it's now up to me; go away and think about it and let me know.

They paid my expenses and the car took me back to Preston Railway Station. I had a cup of tea in the station buffet until it was time for my train. The initial elation at being offered a place on the course and transfer of apprenticeship was beginning to wear off when I boarded the train!

I decided to treat myself to dinner on the train. I was sitting alone at the table, when a stranger came and asked if he could share the table with me; I would think he was maybe in his late 30's early 40's. We got chatting; I told him about my interview and what I was contemplating. Before he got off at Crewe, he said 'I had to do something like you years ago; I expect you will make the right decision'. That phrase has always stuck in my mind. There's no doubt it strengthened my resolve; although as they say nowadays, 'it was a no-brainer'; not if I wanted to get into the drawing office.

When I got back to Saunders Roe, I went to see VT and said I'd like to take the opportunity that English Electric Aviation offered, and wrote to the company accepting their offer. Soon they wrote to me with conformation. They sent me addresses of digs in Lytham. I stayed for a while with Mrs Shorrocks at 31, Church Road, Lytham.

She was kind to me, but being young and having to settle myself into a new area and make friends didn't fully appreciate this at the time. Trouble is when you do finally realise; it's usually all too late. When I went back to visit old haunts a few years ago, the house had been converted into a Doctors Surgery.

Lytham is one end of the Borough of Lytham St, Annes of the famous golf course. St. Annes is the other end. The two towns are separated by about 3 miles of sand dunes; midway is the small inland residential area of Ansdell, where I eventually shared a flat with another draughtsman. About a quarter of a mile from Ansdell on the coast was a big artificial Lake, called Fairhaven. Here you could rent a small sailing dingy and teach yourself to sail with little harm, giving some amusement to passers by! Lytham St. Annes is a genteel Borough in which to live, and a few miles to the bright lights of Blackpool.

In early August after the A1 exams, I left Saunders-Roe and rode my Francis Barnett motorcycle; with my holdall strapped to the petrol tank; up to Lytham, stopping overnight about midway at 'The Swan' Hotel at Lichfield.

I arrived in Lytham next day mid- afternoon after a wet ride. The landlady Mrs Shorrocks was a kindly person, I suppose in her 60's or thereabouts. She'd lost her son in the war. He was an Air-Gunner in Bomber Command. I'd read about him in war books; he'd extinguished a fire in a Wellington Bomber using gloved hands, spent time in hospital recovering from burns for which he was decorated. He was latter killed on his return to operations. His mother and father would never have got over that.

When I arrived at English Electric Aviation, I'd expected to find the sort of Apprentice set up I had known at Saunders Roe; some apprentices in hostel accommodation away from home etc. lads in similar circumstances to myself; ready mates when you met up. It wasn't like that at all; the other apprentices on the course lived at home in Preston or its environs, had mates and girl-friends in their own communities. I don't think it struck any of them to invite me out for an evening; but in any case, Preston was quite some distance away, so it would have been impractical.

In situations when on your own; keep busy; find somewhere to go or do each evening; meet people. So, I joined The Young Conservatives, usually a sociable bunch, the local Sailing Club (The Ribble Cruising Club), The Lytham St. Annes Rambling Club which organised walks in the Lake District once a month; met some interesting people, but not really mates! I also joined the local Amatuer Dramatic Society and appeared in their next production; Desert Song. And explored locally on my bike and by walking.

Once a week I'd walk along the sand dunes to St.Annes to see The Russ Conway Show and take the bus back. Money was tight on apprentices pay. I could only afford a bar of Cadbury's Fruit and Nut for lunch, which I ate walking around the Flight Hanger looking at Canberra, Lightning, Hunter chase aircraft, Dove communications aircraft and a test pilots private Proctor, a single engine canvas and wood monoplane. Those were the days; seventh heaven! Wouldn't be allowed to wander around these days.

Dad sent me the train fare home for my first Christmas away from home.

After I had completed my 'time' as an Apprentice, at the end of January, I received a letter asking me to attend Fulwood Barracks in Preston for a Medical and Assessment regarding National Service. I went to see Mr Daly, the Training Manager who said he would chase up my deferment, the wires had probably got crossed! But in the meantime, I should attend Fulwood Barracks. On offer was the Infantry, or if I signed on for 3 years the chance to go into REME or perhaps the RAF. Bit of a shock, I should have enquired about deferment on arrival at English Electric Aviation, but thought it was covered. Because of day release at Tech. that week I had to complete 5 days course work in 3! Some probably spilled over to the next week, but I caught up, I didn't fall behind; I do remember that!

I attended The Harris Institute Preston on day release to complete A2 for HNC. The day I was at Tech. the Drawing Course Instructor started a new topic. Next day he gave me a potted version and set me to work! He was a mean minded old so and so. I had to do 5 days work in four, and he recommended I went into the Mold Loft! Mr Daly the Training Manager disagreed; he said I should go into the DO as promised.

How did I know? The Training Managers Office was in the next room to the Drawing School and an ex- Navy guy put a set square up against the wall, facing out, with the other end to his ear!! This amplified the vibrations so he could clearly hear the conversation in the Training Manager's Office; trust the Navy; top guy!!

Naturally only having 80% of the time to complete the work the other trainees had for the task, I was one of the last to finish and had to 'hit the ground running'. But I finished all the exercises on time with a high level of draughtsmanship. Sure, the Instructor may have needed to give me more supervision to complete some exercises; but hang on a minute; I'd missed out the vital day when everyone had had the full instruction and the day to get their heads around the new work! I made the lost time up by working through lunch break. I worked hard and deserved my success, but not according to the Drawing Instructor!

He was a crusty old sod. You come across that type sometimes, but usually they have a bit more intelligence and fair mindedness about them. Anyhow, he didn't hold me back.

# 3. DESIGN DRAUGHTSMAN @ ENGLISH ELECTRIC AVIATION LIMITED, WARTON.

The English Electric Company was formed after the Armistice at the end of 1918 by the amalgamation of five businesses involved in producing munitions, armaments and aeroplanes. This new company became one of the country's three principal producers of electrical equipment.

With war in Europe a certainty, English Electric was instructed by the Air Ministry to build a shadow aircraft factory at Salmesbury Aerodrome, near Preston, in Lancashire to build twin engined Handley Page Hampden bombers. By 1942, 770 had been built. A second factory was built on the site and the runway lengthened. Here they constructed over 2000 Handley Page Halifax heavy four engined bombers by the end of the war. That in itself tells you the attrition rate on bombing operations!

At the end of hostilities English Electric began producing de Havilland Vampire Fighters under licence; 1300 were built by the firm.

The company had set up its own design team during the war, initially in a large Garage at Preston, headed by the former Chief Designer at Westland, WEW Petter. Later FW Page took over this role. He was responsible for developing the Lightning and TSR2 Aircraft.

In 1947 the firm moved to the former RAF Warton Aerodrome, near Preston.

The company design team had two major successes; the Canberra twin jet bomber; in service with the RAF from 1951 until the mid 1960's, and the Mach 2 Lightning Supersonic Interceptor. Both designs sold overseas; the Canberra was built in America for the US Airforce by Martin under licence; modified to have a tandum seat cockpit in place of the side by side seat bubble cockpit of the RAF and export machines. The US Air Force designated the Martin variant the B57.

In 1960 English Electric Aviation became a 40% stakeholder in The British Aircraft Corporation. Other members were Vickers Armstrong (40%) and Bristol Aircraft (20%) and sometime later Hunting joined the alliance. This was a Government initiative; the carrot; the TSR2 contract. This was still some years away yet.

When I first joined English Electric Aviation I was placed in the 'Mold Loft' until the Draughtsman's Course started. Here aircraft structural parts were drawn out full size on aluminium plate including bend allowance, holes etc. These were then used to make tooling or reproduced on other plate to be cut out and folded up into parts in the workshops.

After paying my digs, bus fare to work, fuel and garaging for my 'bike', there wasn't much in the 'kitty' for lunch, which was often a bar of Cadburies Fruit and Nut. The aircraft flight hangers were just across the car park from the Mold loft and offices. I used to wander around inside the hangars during the lunch-break as I mentioned previously. On completion of my Apprenticeship on 29th of January 1959, I got a pay rise and my financial situation improved. I could now afford lunch in the canteen.

But before I got organised joining clubs etc. I rode into Blackpool on my bike. I was expecting a sort of Bournemouth of the North resort. The painted steel plate trees, tram lines in the road and trams was a bit of a shock. I quickly parked up and went into the Tower Ballroom to see Reginald Dixon on the mighty Wurlitzer. He came up through the floor playing 'Oh I do like to be beside the seaside', and reckoned he'd got the right job!

The Ballroom was impressive and looked superb, probably had recently been renovated, it certainly was immaculate. You see it on 'Strictly Come Dancing'.

Another time I rode along the coast through Blackpool to the fishing port of Fleetwood. I also got to know the area along the back country roads of the Fylde and inner towns. It was all very picturesque; and coming from the beautiful Isle of Wight that's some compliment.

A regular evening outing that first late summer was a walk from Lytham to St. Annes along the promenade and sand dunes. Russ Conway was starring in a Summer Show in the Theatre on St.Annes Pier, still there I think. It was a cheerful show and the company seemed to enjoy themselves. I caught the bus back to Lytham afterwards. I'd go regularly each week and saw the A and B programs at least twice. The first time I saw one sketch, I laughed along with the audience and actors breaking up when a prop came apart; but this happened again when I saw the same sketch a fortnight later! It was part of the show, but cleverly arranged to look like an accident! That's Show Business. I realised that obvious 'mistakes' are often part of the performance. Real slip ups on stage are usually so well covered up they are difficult to spot unless you are familiar with the production.

In those days the train from Lytham went into the centre of Blackpool; that station is no longer there, can't now get nearer than Squires Gate by train. And then on by tram or bus.

I used to go to dances at the Winter Gardens or Tower Ballrooms on a Saturday night. They had first rate resident bands with about 16 musicians or so. In the Summer Season; Eric Delaney, Ken Macintosh and other 'big names' of the day would take a slot for a couple of weeks or so each, giving the resident bands a break. The resident bands were first rate, giving the 'big bands' a run for their money. Just remember to keep out of Blackpool 'Scots Week' and you'd be OK!

Blackpool's Grand Theatre is a wonderful venue; saw my first Doyle Carte Gilbert & Sullivan shows there. The Mikado and Pirates of Penzance; terrific.' Thomas Round, Donald Adams, Valerie Masterton et. al.

Saw big stars of the day in the Summer Shows; Harry Secombe, Morecombe and Wise, etc. That and exploring a new area, made for a very interesting time.

On completion of my apprenticeship, my salary increased substantially, but until then I subsidised my apprentices pay from Post Office savings if I wanted a night out.

When living away from home for the first time its important to keep occupied and busy. Have a routine and develop interests that aren't going to cost a lot of money.

I rented a Radio, joined the Young Conservatives because they had a good social program and nice people and other clubs to keep occupied. It cost virtually nothing after the subscription. At the Ribble Cruising Club meetings, I made half a pint last the evening. I joined the Lytham St. Annes Operatic Society, where they were working on Desert Song. I appeared as a Riff and Legionaire. There were quite a few changes from one costume to another, so it was important to keep up with the scenes and not lag behind!

The show went on at Easter in the Lowther Pavillion, Lytham. It was situated in a small park, located between Church Road, where I was in digs and the Sea Front with its famous Windmill. So, I stayed in Lytham for Easter that year.

An old long time school friend, living on the 'Island' wanted me to be Best Man at his wedding that Easter at short notice. I didn't have the resources needed for being best man; new suit etc, nor the cost of the journey. I also felt some obligation to the 'Operatic Society' I'd been a member of for some months working on the show. I'd had very little notice of the wedding myself. Had I'd known earlier I would probably have dropped out and bought the financial bullet. But it would have dug deep into my reserves that couldn't be replaced easily at that time. And financially I had to manage for myself; I couldn't ask my parents for help except in extremis; certainly not for something I'd brought on myself. I would also have been letting my newfound friends down. I declined, Nigel was disappointed and never fore gave me! But then he'd never had to live away from home and make out for himself. At that time, I was living payday to payday, I wasn't saving anything, just treading water!

If you have not been placed in that position yourself you don't understand; simple as that.

Eventually the time came to join the Draughtsman's Course. It was run by a long serving employee of the company close to retirement. At the same time, I began attending Day Release at the Harris Institute, Preston studying A2; the final year of the HNC. With one day away at Tech. I had to get 5 days of the Draughtsman's Course class-work into four. It didn't help that the day I was away the instructor started a new topic! Still I kept up and on completion went into the Drawing Office as a Trainee Draughtsman working on the Lightning fighter. The first job's, I was given were handling queries from the shops and modifying drawings, getting to grips with the drawing office system etc. eventually doing design work.

When I joined English Electric Aviation, I somehow thought I would be mingling and socialising with their apprentices. But it wasn't like Saunders-Roe with perhaps nearly half the apprentice intake from the mainland, hostel based and consequently outward going and socially inclined. The English Electric Aviation apprentices lived at home in Preston, they had their own social network; not necessarily a close association with other apprentices; some had girl friends. It was a different set up; they just got on with their lives as they had always done. Outsiders just didn't register on the radar, nothing personal, that's how it was.

At Saunders-Roe there was an intermingling of graduates with apprentices as they followed more or less the apprentice's progress through the shops. But you get on with it and make your own local circle of associates and friends by doing as I did; join local clubs. It certainly makes you independent, no bad thing. But you are starting without a nucleus of friends or aquaintances from school etc.; so, you really do have to make out for yourself. As Dad said to me, keep out of the pubs and away from dodgy characters. Keep occupied is the answer, be outward going. It worked and I made a good circle of friends.

I was shown kindnesses by various people; not least by an oldish Jewish couple in particular who ran an 8 till late café in Lytham, in Church Road, not far from my digs. They used to see me go by with a parcel to post (my Laundry home!). The lady stopped me one day and said, why don't you drop your parcel off here on your way to work and I'll go and post it for you, pay me when you come home from work. I thought that was very kind. I knew they would have been pleased to befriend and make a fuss of me. But I wanted to remain independent, silly boy really. They were probably of grandparents age. Always got a wave when I passed by. It made a difference.

When parachuting into a new area and; to some extent culture; at age 20 or so, it's not as easy as the 'Agony Aunts' would have you believe to make friends. Your not in a college Campus rubbing shoulders with perhaps hundreds of like minded youngsters; your on your own. You don't even half know anyone to wave to. You know no one! Just making a point! You do what I did, get out and meet people, join clubs etc. If you are good at sport then there is tennis, judo, and so on. If your less than average at these activities you stand out as a bit of a fool!

I developed an interest in HI-Fi; big at the time. I would go into the Hi Fi shops in St.Annes; Northern Radio Services was a favourite and talk to the salesman; probably ad nauseum, but it was someone to talk to! I was given a ticket to attend the Audio Fair at Harrogate. I rode my Francis Barnet motorcycle over the moors to Harrogate. Leak and Quad with their Electro-static speakers were some of the top brands in those days; very expensive well beyond my price bracket, but interesting. I got an earful of everything going and came back with lots of brochures. I said to Mrs Shorrocks, I was thinking of getting a Hi Fi system; would she mind. No that was OK; then I dropped my bombshell! My uncle's wife had come from the North originally and now lived in Chichester. She said to me before going up to Lytham, people up North are straight talkers, say what you mean; I always did, so didn't really need encouragement in this respect!

Taking this advice on board, I then said to Mrs Shorrocks, does your furniture have woodworm! She gave a short, startled laugh and said why? Here came the punch line; I said because it looks the sort of furniture that might have! Well… one of the other lodgers said shortly after, if anyone had come looking for lodgings just then, you would have been out!! I don't think we quite had the same relationship afterwards. Problem was, I didn't think I'd said anything wrong. Eventually I sold my bike and bought some Hi- Fi of sorts! Two mistakes for the price of one really!

I found concentrating on Technical College work difficult away from home. I didn't have the same facilities for study; my bedroom was unheated and cold in the Winter; I didn't have a desk or chair; the lighting was inadequate. I suppose I could have done something about it if I had approached Mrs Shorrocks the right way, but I was young. The digs weren't set up for study, and to get what I wanted would have put the weekly cost of digs up!

The whole set up and situation was new to me, so the studies suffered. It wasn't helped that a fortnight before the exams my landlady announced she was going on holiday.

She found some other digs for me, but I had to share a bedroom. I couldn't get at my books, or study, it was unsettling so I gave up. One of those things! It set me back a year, but there it is. I had a lot on my plate and didn't want to be a recluse, I needed to get out and about, meet people etc. Maybe if I was being mentored it would have been different; but that's how it was. If it was difficult studying in the digs, you were up against it. No excuses; that's life! Too much on my plate.

During the Summer, my parents travelled up to Lytham for a week's holiday. Mrs Shorrocks provided bed breakfast and evening meal for them. I'd have evening meal with them. I took some time off work and rented a car for two days; driving them around the Lakes one day, and another time down to Chester.

Other lodgers came to stay from time to time. One I remember was Graeme Elkington, a Flight Test Observer. He was later to eject from a Lightning trainer; eventually leaving English Electric to work in the United States. He would come down to breakfast and grab the toast, making marmalade sarnies for lunch. I got a lift with him one evening; can't remember where; but he had a disconcerting habit of driving at oncoming cars that had headlights on full beam! Another lodger Bob had served in the RAF as a Technician; he had transferred to the Kiwi Air Force (RNZAF) before training at English Electric servicing Lightnings, after which he was going to Saudi to make his fortune; interesting people. Wouldn't meet them living at home!

Before Tech. started again, I had decided to find a flat, where I could settle down comfortably and do tech. work; digs weren't suitable and I didn't want to risk another lock out!

David Arkwright; another draughtsman; was also looking for a flat; so, we joined forces and moved into accommodation at Ansdell. Mrs Shorrocks was upset when I gave notice; one of the other lodgers said she thought I'd moved because my parents told me to! Nothing further from the truth; again, I should have told her again why I was leaving; had I tried she like as not wouldn't have believed me. But I should have made the effort.

Later on, I heard she had moved with her husband to live in Tunbridge Wells to be near her younger son who was Town Hall Clerk there. I hope it worked out for her.

After I transferred into the DO from the Draughtsman's Course, a draughtsman introduced me to George Carr who ran the local Lytham St. Annes Rambling Club. It was a good samaritan who introduced me to George.

The club travelled by Ribble Leyland Coach to the Lakes once a month for a ramble. I watched the sunrise from Helvellyn once, when we travelled up on a Saturday afternoon, spent the night on Helvellyn to watch the sun rise, then down to the car park where the girls cooked breakfast over a wood fire, followed by a Lakeland ramble and then home. Super weekend.

I went to the Harris Institute next year to sign on for A2 again, but this time evening classes; three nights a week. The senior lecturer in charge of enrolement said he was disappointed in me, and that I should first repeat A1 before doing A2. I replied you want to try living 300miles from home on your own knowing no one and two weeks before the exam the landlady goes off on holiday and you can't get at your books or study. See how you get on! He backed down and let me enrole on the A2 course.

Two other draughtsmen at work were also doing the same A2 course. They lived in St.Annes; along the coast from Ansdell (midway between Lytham and St. Annes) where I was then flatting so I got a lift from work to Technical College and then back home. We shared costs. The firm were very decent; if we worked after 6-30p.m. doing overtime they provided a free tea in the canteen. It may have been a meat pie with bread and butter and jam, with a cup of tea; something like that. Whatever, it was very considerate of the company and carried me over till I got back home from classes at nearly 10 p.m.

My flatmate was another draughtsman, David Arkwright, about 4 years older than me. Rather than doing 18 months National Service after his apprenticeship, David joined REME for 3 years and came out a Sargent; pretty good going; a really decent bloke.

He was good at sport, a no- nonsense sort of person and a good friend. I guess he thought me a bit wet behind the ears. My father took to him when I took him home on holiday. Looking back, I think he looked after me as a younger brother. Sadly we lost touch. However, quite recently we have been able to get in touch again through his son. Long story, won't go into it here.

David and I flew to Ireland for a holiday. Stayed in Dublin a couple of nights, rented a car, drove across to Galway, where Joe Loss and his band were there for the races; then we drove down to Killarney;

a lovely spot; then across and to Cork; which didn't appeal; and back through Waterford to Dublin.

Skerries, was a small seaside resort not far from Dublin; we went there to a couple of dances. Boy don't the Irish know how to have fun! Wow.

Another time I was driven around Scotland by another mate, so a very good time all in all. But I digress.

The evening classes I recall were on Tuesday, Wednesday and Friday. Thursday, I tried to clear as much of Tuesday and Wednesday's homework as I could. The 'golden rule' was to get that week's work completed before next week started. It was important to have cleared all that weeks evening class work by Sunday night' otherwise you were in danger of getting over-whelmed.

That could mean staying in on Saturday if you were really loaded down with Lab Work as well as homework. Lab work, I remember, used to come in pairs; if one subject had lab work that week you could be sure another subject would too.

Getting behind with work was lethal; you had to keep going; and keep your head above water. It was hard work and tiring. Sometimes I went to bed early on a Saturday rather than go out. I was working at my limit; but I got decent marks second try. I was quite proud of myself. And I was a Design Draughtsman.

It's worth mentioning that the Higher National Certificate subjects were at degree level; especially hard when not studying in ideal conditions. I think it's fair to say that had I been able to remain at Saunders-Roe and at home, I would have sailed through HNC first try and taken the endorsements in my stride. But I would have lost out in personal development and wouldn't have met my wife. I made the right choice and became a more balanced and capable person for the experience.

What of girlfriends? There wasn't time to get involved when busy studying; you just can't afford diversions if you are going to pass the exams. However, if the right girl had come along during that time and she was interested, I would have managed somehow, but not a risk I wanted to take, or had time for during term time; too much at stake; not at that time in my life.

I got to know girls at dances over the weeks and sometimes went out a few times together; that sort of thing, but nothing serious. I preferred the older girl; I think that's common with young men, but the girls were interested in settling down. I was looking for friendship; I didn't want to get involved. We'd only go out a few times; which is enough unless you really hit it off. When you do, you'll know all about it! Not love at first sight necessarily, but you'll almost at once, think, that's the girl for you, even if there's a risk it may end in tears!

These friendships gave me the experience to recognise when I did meet that special girl. I had to wait for it, but when I did, I realised I'd better not mess about, because they clearly don't come along very often., or stay unattached for long. Faint heart never won fair lady; very true. Again, you make your own Good Luck.

The year after the HNC evening classes, towards Christmas, the 'Rambling Club' committee organised a weekend and dinner in the Lakes; staying at The Old Dungeon Gyll in Langdale. We shared cars. I was taken up on Friday evening after tea. It was a couple of hours drive so we arrived with drinking time to spare; I hasten to add 2 pints was my limit. After breakfast next day I went rock climbing; a 'first for me'. To my surprise I enjoyed it. It was an easy climb of course behind 'the gyll', but it certainly boosted my confidence. I was roped up, 'just in case' and absailed down; a great experience. It put everything else in its place. Shyness disappeared! On the Saturday evening we had the Dinner; a very enjoyable evening. Sunday, we went for a ramble and then home. They were a good bunch of people.

That year, after the marathon 3 nighter of the previous year, I'd started studying Heat Engines one evening a week at the Harris Institute; can't remember how I travelled back and forth. We'd been pilled high with homework over Christmas; it over-faced me; I dropped out; the previous A2 year was very hard work, hadn't the stomach for another flog!

I was on such a 'high' after my climbing, I just went up to a girl I liked the look of in the local bread shop and invited her to the DO Christmas Dinner Dance. She came, but somewhat lost interest; we didn't have a lot in common. Sometimes happens on first dates, I didn't persue it. Anyhow after she turned off, I let her mope and went and enjoyed myself.

There were three others with London and Bournemouth roots in the design office, so at Christmas time we would club together and hire a big car and drive down with drop offs in London (don't remember where); with me at Southampton. The driver took the car to Bournemouth and had use of it over the break. In those days Christmas was a short holiday; we'd finish mid-day Christmas Eve and travel back up North after Boxing Day. A much better way of travelling than my first Christmas journey home; the London train came into Preston Station pulled by two locos bringing it down from Windermere. Navy ratings were travelling down to London in the same compartment; their technique was to have a few bevvies and sleep the journey away!

When the TSR2 Bomber design was started, a contingent of Engineers from the engine supplier Bristol Siddeley Engines used to fly up from Bristol every so often to design meetings. They walked along the side of the DO. I noticed one man in particular, he was big and immaculately turned out. I later worked for him in the Bristol Siddeley Engines stress office some years later; small world. I found out then his name was Bernard Grant.

I continued working on the Lightening F3.

The DO then comprised: Chief Draughtsman; Group Leaders who had a couple of Section Leaders, and under them Senior, Intermediate and Junior Draughtsmen. I was in the Junior category. Each Group would have a Checker. He was a Senior Draughtsman of section leader status, whose job it was to check all drawings for dimensional accuracy, clear unambiguous notes, the drawing schedule of parts, specifications and check for any other errors or mistakes.

Our Group Leader was Bill Marrs, famous for playing a trumpet in the DO at the break up for Christmas in his younger days; he was tall lean and probably then in his 60's; Conrad Holtapple, a Dutchman was his right- hand man.

Eric Talbot was our Checker. He was fastidious and scrupulously read all notes to see if he could find ambiguity or lack of clarity. He would mis-read the notes; if he could. It drove some draughtsmen up the wall; but I welcomed it. It was excellent training and I think probably the reason I usually write clearly now.

Other Draughtsman's names that come to mind are: Ron Cook, Dennis Liddle, Arthur Parkinson (or Parkman), David Arkwright, Norman Lee, Jim Whigston, George Carr… I remember the faces, but the names fade.

On completion of my Draughtsman's Traineeship, I was paid Junior Draughtsman's rates. I had enough money in my pocket to eventually buy an old car. A 1938 Morris 12, bit of a rust bucket, but good mechanically. Running that took my spare money. I used it to drive David and myself to and from work. He kept his Triumph TR2 sports car for Sunday best.

I drove my old car up to the Lakes one day, and in a few seconds inattention clouted the front wing along a stone wall! It happens when your new at driving, I'd just thrown an apple corps out of the window and handn't noticed a turn in the road. Water was seriously leaking out of the radiator, I straightened up the torn wing, drove to a garage and fixed the leak with Radweld; filled up the radiator and drove back to Ansdell where I was living. With the torn wing I drove to the Library in Lytham. The Library was close to the Police Station. As I came out of the Library I saw a copper crossing the road. A few days later I received a summons…' I was apprehended whilst driving along this road for driving a car in a dangerous condition….' and so on along these lines. It was really made out to something more than it was, and it was untrue in that I hadn't been apprehended!  I spoke to my Section Leader at work about it and showed him the summons.  Remember it was in the days of 'Dixon of Dock Green' TV series, when the Police were painted as kindly and good hearted.  My Section Leader said, take this down to the Police Station and see the Sargent, he'll probably quash it; that was the image of the police at the time.

The Sargent said how is this wrong siiir?  I explained; he said leave it with me.  In due course I got another summons, still not right, but the 'being apprehended' bit had been left out.  I was fined.

As soon as I realised, after the first summons, the car wing needed tidying up and tears smoothed out, I'd straightened everything out and fibre-glassed over it to make a decent repair; not a professional job, but safe.  Good at roundabouts because everyone gave you priority!

A while after I'd paid my fine, I came out of the flat and walked round the back alley, where I kept my car; and who should I see but the same copper who had booked me taking a keen interest in the repair!  Didn't he jump when he saw me; clearly after a supplementary conviction, perhaps getting back at me for complaining about his original summons.  I could have probably reported that copper for harassment; which is what probably worried him. Wouldn't have endeared me with the local force I suspect.  Just shows how a 'rotten apple' can undo all the good public relations work of 'PC Dixon'.

After 4 years at English Electric; or British Aircraft Corporation as it was then, I decided I wanted to get nearer home.  Lancashire was too far way.  I didn't want to live at home, not after being used to my independence.  It would also have been difficult to leave again, but really there was no question of going back home anyway because employment on the Island was unstable.  British Hovercraft Corporation (Saunders-Roe) the natural choice was at best just ticking over and had its ups and downs.

And when you have won your independence don't want to let it go. Take a girl out a couple of times and mum would want you to bring her home for tea; stay out late too often and mum would get dad to have a word! That sort of thing. And then of course all my apprentice friends had moved on and left the Island.

I drove the old Morris down, packed with my stuff, I'd aquired, down South on November 5$^{th}$. I remember the drive very well with fireworks going off in the towns I drove through. I got to my Uncle and Aunts house in Chichester about 9-30pm unannounced and stayed the night, continuing my journey down to the Ise of Wight in the morning.

# 4. DESIGN DRAUGHTSMAN @ F.G.MILES ENGINEERING, SHOREHAM BY SEA

F.G. Miles interest in aviation began in the early 1920's with aircraft repair, overhaul and reconstruction work at what is now Shoreham Airport.

Local pilot Cecil Pashley became a partner in the new aviation business and taught Miles to fly in Pashley's Avro 504K. FG went on to become a flying instructor and take people up for joy rides.

In 1928 F.G. Miles designed and built the Martlet biplane trainer of wooden construction. Six were built. He followed this with the Metal Martlet; only one flew.

He was financially supported in his ventures from time to time by his father who had a Laundry business further along the coast in Portslade.

Miles next design was the M1 Satyr; built in the Parnall Aircraft works at Yate, Bristol. Flying the aircraft from Yate to show his parents, he landed at Reading Aerodrome for lunch. There he met Charles Povis of the firm Phillips and Povis. They got into discussions on a future light monoplane. It was soon agreed that Miles Aeroplanes would in future be built by Phillips and Povis at Reading. The aircraft would bear the Miles name. The first aeroplane of this collaboration was the two seat M2 Hawk; a tourer and training aeroplane of which 55 were built.

In 1935 Phillips and Powis became a public company with FG Miles as Technical Director. In 1936 Rolls Royce took an interest in the company and FG became Managing Director.

In 1941 F. G. Miles took over the Rolls Royce interest and in 1943 the company name was changed to F.G. Miles Aircraft. The M2 Hawk Major, was a development of the Hawk. A prodigious number of new designs followed apace including the: M3 Falcon. a 3 – 4 seat cabin monoplane; M4 Merlin, a 5 seat cabin monoplane; M5 Sparrowhawk, a more powerful hawk development; M6 Hawcon, a 2 seat experimental thick wing aircraft; M7 Nighthawk, a development of the Falcon; M8 Peregrine twin engined 6 – 8 passenger transport; M9 Kestrel advanced military trainer and target towing aircraft; M11 Whitney Straight two seat side by side cabin monoplane; M12 Mohawk, dual control tandem two seat high performance long range cabin monoplane, built in 1936 to the specification of Col. Charles Lindbergh; M 13 small single seat racing/research aircraft, M 14 Magister two seat elementary RAF training aircraft known as the Maggie; M15 two seat elementary trainer; M 16 Mentor two/three seat dual controlled light cabin monoplane; M17 Monarch a development of the M11; M 18 intended as a Magister replacement; M20 single seat twelve gun fighter designed in 1940, intended as a utility fighter replacement in the Battle of Britain, but not needed; M25 Martinet and radio controlled M50 Queen Martinet two seat target towing aircraft; M28 built during 1942 as a private venture, three or four seat cabin monoplane, six were built; M30 X Minor a two seat experimental cabin monoplane investigating wing/fuselage interaction aerodynamics; M 33 Monitor high speed target tug with crew of two; M35 Libellula private venture tandem wing research aircraft, would be called a canard configuration today; M 37 trainer version of the Martinet target tug; M38 Messenger, a development of the M28 Mercury; M 39B Libellula, two seat tandem wing (canard) twin engined research aircraft, intended to investigate a future bomber design. The M 52 high speed research aircraft with a top speed 1000 mph was cancelled in 1946. This concept and technology was given to the USA, amongst the innovative ideas at the time was the all flying tailplane, which was a key element of control in supersonic flight. An independant assessment of this project a few years ago concluded that if built it would have performed as predicted.

The M57 Aerovan was a light twin engined short haul freighter, 39 were built; M 60 Marathon first four engined all metal aircraft built by Miles. This project caused the collapse of the company due to Ministry delays. Handley Page took over the contract at a price greater than that which Miles agreed to build them for! Forty of the type were built; M 64 LRS two seat light monoplane for the club/private owner built by Miles employees 1944/45; M 65 Gemini four seat twin cabin monoplane, 170 were built; M 68 Boxcar four engined short range freighter; M71 Merchantman, all metal larger version of the Boxcar; M75 Aeries development of the M65 Gemini; M 77 Sparrowjet, the prototype M 5 Sparrowhawk was much modified and fitted with two Turbomeca Palas 330 lb static thrust jet engines in 1953. It was destroyed engineless in a hangar fire in 1964; M 100 Student two seat side by seat all metal jet trainer, developed as a private venture. It first flew in 1964. It was not proceeded with further; the RAF bought the Jet Provost, South Africa were interested in the Student, but there was an arms embargo at the time; HDM 106 Huriel Dubois high aspect ratio wing fitted to the M57 Aerovan; M114 and M118 two/four seat light touring aircraft and developed under the auspices of Beagle.

In 1943 Miles saw a prototype ballpoint pen made by Laszio Biro. Miles put this into production and formed the Miles Martin pen company after the war.

In 1943, Miles and his wife Blossom; a mathematician, draughtsman and stress engineer formed the Miles Technical School

In 1949 some years after the company at Reading had collapsed, F. G. Miles moved back to Shoreham Airport where he first started, to set up a new company. In 1961 the aviation interests of this company became part of BEAGLE. Across the River Adur from the Airport, Miles built a new premises for two companies; F G Miles Engineering and F G Miles Electronics.

In 1975, Hunting Associated Industries acquired a controlling interest in F G Miles and all its subsidaries.

Miles lived in Cudlow House, Rustingdon, the inspiration for J M Barries Peter Pan. Miles died on 15$^{th}$ August 1976 at Worthing Sussex.

I joined F G Miles Engineering in 1962 from English Electric Aviation as a design draughtsman. The Chief Engineer was John Parker, the Chief Draughtsman Ken Wells.

Short Bros. developed the Aerovan concept into the Skyvan. F.G. Miles Engineering had the job of designing the electro-mechanical Flap Actuator. The company were also developing a range of commercial electro mechanical actuators; anti-collision aircraft beacons and electric motors. Other work included the design of an Actuator test rig, modifying US bomb release equipment for installation in V bombers; trial installation of equipment in armoured vehicles at Chertsy. And design of Nuclear Analytical Equipment. Nothing mind-shattering, but a broadening experience.

F G Miles Electronics; a parent company on the same site; were designing a Flight Simulator for the Fleet Air Arm's Buccaneer fighter bomber.

I used the old Morris to commute and drive up to Chichester to see my relatives during the week sometimes. One evening on a misty evening I saw a Vulcan Bomber land at Tangmere; in the misty conditions it looked like a giant ray.

For a couple of week's the firm put me up at Miss Menzies Guest House in Shoreham; very nice place. Also staying there was a Personnel Manager relocating into the area. He was an ex- Royal Navy Commander who had taken part in the Battle of the River Plate in WW2. The film was on release at that time, so an interesting topic of conversation.

I then moved into digs where the husband and wife really didn't get on, the wife was asthmatic and chained smoked; the husband miserable, but OK to me.

Didn't stay long and found new digs around the corner with a retired brother and sister. He had sold Insurance, she had been Manageress in a large London store on Oxford Street; Debenham's I think. They also had a small dog; a Pekenise, who was a bit possessive and I think jealous of me! On going to bed one evening I found his visiting card on the eiderdown. I told my landlady the next morning and she said she would sort it. Next night there was this smell! I told her next morning and she said well I did put the Eiderdown out to air!! You meet them all!!

Not far from the digs in Old Shorehan Road was the Red Lion, close to the old road bridge across the Adur. Cosy snug, big fire, pint of Watney's Red Barrell; what more could you ask for.

One weekend I brought the Morris home and took the engine and gearbox out, keeping that in case I bought a boat! Sold the rest of the car for scrap; it was sounding a bit creaky and thought it may not last much longer.

I used to visit the car dealers around Shoreham and Portslade looking for something suitable to get me about. An Austin A30 would have done the job. I gave my Shoreham address to a car salesman in Portslade, a town situated between Shoreham and Brighton. One evening the salesman called at my digs and said an A40 Sports had just come in; was I interested! I didn't know what that was, but it sounded interesting. He took me to see it and I bought it.

When I was 18, I hankered after an MG TA two- seater sports car. They were regularly advertised in Motor Sport for about £275; a small fortune to me on apprentice wages. But when I was living in Shoreham, I saw an MG TC (even better) advertised. I went to see it and had a ride. It looked in fine condition, but the ride was a bit sloppy; seemed like it needed new shock absorbers. The owner wanted too much for it, so I walked away. The A40 Sports, although not iconic was a better car, very comfortable and went where you pointed it! Comfortable and great on long journeys.

Went to see VT on one of my home visits and appraised him of my doings. He was pleased to see me. I was enthusiastic about my new job; VT said now 'the worlds your oyster' Peter. And so it turned out to be!

On another of my visits home I went to a dance at The Trouville Hotel, Sandown; one of our old apprentice haunts and met up with a bunch of ex apprentices who had congregated there that night for a get together.

There I met up with Brian Simmonds who was captain of Worthing Rowing Club; he now lived back home in Littlehampton. He invited me over to Worthing to go rowing. That was the beginning of a very good couple of Summer's. Once a fortnight we'd put the boats on top of a South Down coach and travelled the south coast to the regattas. I'd sometimes turn out for a game of coarse rugby in the winter!

We would train twice a week on the sea. Afterwards repair to the Frog Pond snug for a pint, and undo all the good work.

Apart from the South Coast regattas took part in head of the river race at Bideford Devon sleeping in the clubhouse. There was also the Rowing Championships on the Surpentine lake in Hyde Park; a day trip on back of a lorry with the boats from Worthing! It was good fun, but didn't win; it was the taking part!

On the journey back to Worthing, I wanted a pee; the lorry stopped by a Public Convenience, I hoped off the lorry to relieve myself. When I came out the lorry was gone! I flagged down the Portsmouth Rowing Club Coach and got a lift back to Worthing; a very jolly journey indeed, Pub stop and sing song included! Not sure the locals were too impressed, but the Landlord did good business that evening.

I bought a 1952 Austin A40 Sports. It was bodied by Jenson with aluminium panels on an Austin Devon chassis with a 1200cc engine with twin carbs; column gear- change. Beige with red leather (MOV 732). Quite stylish with very comfortable leather seats.

I drove it up to Lytham to see my old flat mate one weekend and had a day out in the Lakes; we took out his girl- friend and the sister of a girl I used to know. I thought it would be interesting to renew her acquaintance only to find she was engaged, so the sister came instead. The lesson there was clear, which I put to good use when I met Sally; I didn't let an opportunity slip through my fingers again.

Seemed to be able to go on for ever in those days; drive everywhere, do anything, and not get tired; not now though!

A personality I worked with in the DO at FG Miles was Ken Ritchie who had worked in the Donald Campbell team developing the Bluebird car and jet boat. His had been an interesting life. He'd come back to settle in Shoreham by Sea with his family.

I fitted new rings to the pistons and also big end shells to the A40 Sports and took it on the Continent with a rowing mate; driving down to Geneva, Montreau, Milan, Genoa and along the Mediterranean coast through Nice, Monaco to St. Tropez. Then up to Paris and home. Quite a trip. My mate was organising a tent; he turned up with a one- man tent; not impressed; he was a bit of a wide boy I thought. Either I slept under the stars or in the car, once when the weather was bad; we stayed in a hotel.

On the last lap in France near the ferry, the car suddenly spun on a slippery hill, came to a stop close to a lorry! Driver gave sign of the cross and smiled. Car had a broken rear spring which may have broken on braking causing the spin. Phoned a garage. Black Traction Citroen (shades of inspector Maigrait!) came out, driver indicated it would be OK if I drove carefully. He did this by sign language mainly; that was the extent of my French; not proud of that but how it was. Had both rear springs replaced when I got back to Worthing where I lived in a bedsit.

At work I wanted more of a technical challenge than I was getting; so, started looking around again. In those days Technical College Teaching courses were widely advertised. I thought I'd try. I got an interview, but it got no further than that. I don't think I would have been suited at that stage of my career anyway.

I later saw an advertisement for a Weights Engineer at Dowty-Rotol, Cheltenham. It sounded an interesting job; estimating weight, centres of gravity and moments of inertia of undercarriages;

estimating jack loads, weight control and basic stressing etc. That seemed the ideal next move for me, technical challenge, excellent company and super part of the country. I applied and got the job. I found it easy to find a job in those days; I suppose my CV was fairly impressive for a young draughtsman. The feeling of the time was to give you a go if thought suitable and train you up. Bit different now; many firms are as good as the current contract, with not much work to bid for. They want people who can hit the ground running!

Not long after I left FG Miles, they won the contract to design and build the replica aircraft for the film 'Those Magnificent Men in their Flying Machines', some interesting work there!

# 5. WEIGHTS ENGINEER @ DOWTY ROTOL LIMITED, CHELTENHAM.

In 1935, George Dowty set up his own company in Cheltenham making aircraft equipment. When World War 2 broke out, the company was known as Dowty Aviation, designing and manufacturing aircraft hydraulic systems.

Rotol Airscrews; a joint Rolls Royce Bristol company; was acquired by Dowty in 1960, who also acquired Bolton Paul Aircraft Ltd in 1961.

For a number of years Dowty Group collaborated with the French Group Messier on landing gear and hydraulic systems.

In the 1960's the Dowty Group developed fuel control systems and undercarriage for the Harrier jump jet. The company also developed flying control systems for Concorde, and was busy on TSR2 hydraulic equipments.

Undercarriages were also designed for the Fokker F28, and the nose landing gear for the RN Phantom. At that time the company was heavily involved with TSR2, P1154 (Harrier supersonic replacement), AW 681, Battlefield STOL Hercules size transport, all cancelled when Labour came to power. Almost Concorde as well, but for the French!

In the late 1980's the company diversified into telecommunications and computer equipment. It acquired Datatel (software) in 1987; Case Group plc (telecoms) in 1988 and Dataco (network provider) in 1988.

Dowty Group was acquired by T I Group in 1992. In 1993, T I Group hived off seven former Dowty electronic companies in a management buy-out, forming Ultra Electronics.

The Dowty Landing Gear business was transferred into a joint venture with the French company SNECMA, and known as Messier-Dowty.

I started at Dowty Rotol shortly before Christmas 1963. I remember having to put the start date back a week because I caught flu.

The Chief Weights Engineer was Gus Pitt; a gifted Engineer and super bloke. I kept in touch with him until he died. When he visited Auckland in 1980 to visit friends he flew down to Wellington to see me. I was living in rented accommodation at the time, serving notice before I returned to the UK with my family. We had a standing offer to go and stay with him in the US, but for one reason or another never managed it. A lost opportunity.

I palled up with Ken Keyes, another young Weights Engineer on the section. A very friendly person; it wasn't long before I was intermingling with his friends other Dowty-Rotol ex-apprentices and graduates of my age. Something I'd expected at Warton, but an unexpected bonus in Cheltenham. It all augured very well. Really were a great bunch; and the work was interesting too. Gus was a great boss, after I left we corresponded until his death. Should probably have stayed, but then would have missed out on other adventures.

Dowty Rotol had a hairdresser on site and you were allowed to go and get your hair cut in firms time; you booked your slot first of course. BBC Radio were still broadcasting Workers Playtime. One day a show was broadcast from the canteen, starring Ken Dodd!

Our Section Leader, John Bell, had serviced Churchill's transport aircraft during the war; we nicknamed him squadron leader Bell! He was the quintessential Englishman; he later emigrated to Australia. A very mild mannered careful man; I hope all went well for him.

I was having a great social life; went out with the Gloucester Mountaineering Club a couple of times; once to an ex- miners old cottage in South Wales they were renovating. Saturday was work on the cottage; Sunday Rock Climbing and home. On another occasion I went rock climbing in Derbyshire. I also joined the Gloucester Rowing Club, rowed on the canal and went to a couple of local regattas. Friday night was Pub night on Cheltenham Prom with the lads.

I often went to dances at Cheltenham Town Hall. Good local bands, but not up to Blackpool standards. One Saturday night I danced with Sally; who is now my wife. I'd first danced with her blond cousin. As I asked her cousin to dance I caught a glimpse of Sally and I thought wow! Just like that. I got the next dance from Sally and we remained together all evening. She was at Teacher Training College at Newton St. Loe just outside of Bath on the Bristol Road That weekend she'd travelled up to Cheltenham to meet her mum and see her granny. Thinking fast I said 'I go down to Bath sometimes'; a total fib; how about we meet up next Saturday and go for a meal. Which is what we did. From then on I travelled down to Bath on Saturday and stayed Bed & Breakfast in a small cottage in Newton St. Loe. The village was used to putting up parents and friends of students. Sally made the first booking for me. And so it went on. Bristol, Bath and Cheddar is a wonderful playground.

In those days Bernie Inns were all over the area. The company leased a number of old pubs and inns, refurbishing them sympathetically and with taste. All smart, clean and fresh. The waitresses wore black dresses and white pinafores. First you would enter the bar, buy a drink and order a meal. When a table became vacant you were taken into the dining room, where you would probably have a glass of wine; maybe a bottle; or something else to drink. After the meal you were moved onto another bar where you would have coffee, maybe with a liquer. No drink driving laws then; if you felt OK you would drive; 2 pints or equivalent was about my limit!

Berni had 4 basic menus: egg omelette, chips, peas and garnish; plaice and chips, peas and garnish; gammon, egg, chips, peas and garnish; steak, chips, peas and garnish, followed by special icecream or cheese and biscuits. It also included a crusty roll and butter with the main meal. Drinks and coffee were of course extra. The cost: 5

shillings (25p), seven shillings and six pence (37.5p), ten shillings (50p) and twelve shillings and sixpence (65.5p) respectively. I think those were the prices, but you have to remember that before tax and other stoppages my pay was about £19 a week. About £4 would give you and your girlfriend a good night out. The Berni schooners of Port and Sherry were really good. I suppose being based in Bristol they had the connections with the fortified wines trade; Harvey's for instance. What would it cost today? Probably around £50. Digs and courting pretty much used up my wages.

At Dowty-Rotol, I was called Cas. After Cassanova after it got out I was going down to Bath weekends to see my girlfriend! All was fine until the Labour Government getting elected on the White Heat of Technology ticket cancelled; TSR2, the Supersonic Harrier Project (P1154) and it's engine (BS100), and down the road at Gloster Aircraft Division of Hawker Siddley, the AW681, short take of and land battlefield transport. It was about the size of a Hercules, had swept wings, used the Pegasus rotating nozzle Harrier engine. Crucially Dowty-Rotol had a lot of design and future production work depending on these projects.

Development of the Harrier continued. It was sold to the RAF, and eventually the RN for use on their small aircraft carriers (through deck cruisers). The US Marines flew the Harrier as did the Spanish Navy. It was produced in the US for their Marines.

In place of the cancellations, the Government bought the US McDonnell Phantom, but with a high UK content; Rolls Royce Spey engines in place of American engines; a big expensive Engineering task. Dowty-Rotol supplied the Nose Landing Gear. I was involved in the weight estimates of a new design. Why a new nose landing gear? I think it was something to do with increasing angle of attack of the wing to the airflow and shorter take-off on a ships catapult say. I don't really know.

No redundancies were announced, but to me the writing was on the wall. I was serious about Sally. If I lost my job at Dowty-Rotol, I couldn't see much on offer close to Bath. I hadn't connected with Bristol for employment; I should have, but I didn't.

I think my boss Gus Pitt was dismayed at the decimation of the UK Aerospace Industry by these Government cut-backs. Not long after I left, he went to Boeing designing the Inter- continental Missile Silo damping system to protect the missiles from a pre-emptive nuclear strike. And worked on Boeing Airliner Landing Gears; a clever man.

# 6. SHIP DESIGN DRAUGHTSMAN @ MOD, BATH.

Whilst wondering about job security at Dowty's, I saw an MOD Ship Design advertisement for Bath. If there were redundancies at Dowty, being a newcomer, thought I'd be vulnerable.

The MOD advert. was for Draughtsmen. I thought it would be a step backwards, but reckoned I'd soon move myself back up. How nieve; in the MOD! In the Civil Service as with any large organisation, it's particularly difficult to move up a job category or strata. You would at least need an influential sponsor and be outstanding. But the MOD seemed secure and got me into the area I wanted to be; so, took the job. It could have been a job for life in a lovely area; it could have worked out. But had I stayed; in not so many years later when the Ship Design Department was closed down, there were massive redundancies. So, I would have been caught one way or another. Engineering can be a precarious profession!

It was the most boring place I'd ever worked. I was given a pile of drawings to check; made suggestions on how the design could be simplified and cheapened off. Gave the results to my Section Leader and he said 'All that has been checked and signed off by another section, all you have to do is check that the paint scheme called up is Battleship Grey, Tint C'! I hope I haven't given any national secrets away!!

The lesson; don't take a job of lower status or category than you already have. Unless you move that will be the category you will be labelled with in that firm and by future employers. They will probably wonder why you down graded yourself. Think perhaps you were out of your depth before. Another lesson!

Soon afterwards I saw an advertisement by Bristol Siddeley Engines at Patchway on the outskirts of Bristol for a support engineer for their Aircraft Project Analysis Group involving Aircraft Layout and

Weight Estimates. I got an interview and was offered the job. I asked that if a Stress Course came up at BAC Filton; which in those days they did from time to time; could I go on it. This was agreed verbally; should have got it in writing, but the firm kept to their word; I was lucky.

So, another rule; whatever has been offered to you at interview, or you think has; when you get the offer; if any special verbal agreements you understood were made to you aren't mentioned; write back, say what you thought was understood, and ask for clarification. If they are not offering the opportunities you thought were on offer, you can still accept the job or decline. The important thing is you know where you stand; no disappointment or upset later. You may decide to stay where you are; maybe use the offer to try and improve salary, opportunities etc. in your current firm. That is sometimes the only way you can get a salary rise. But you will probably be levelled down again over time! My boss at Vosper Thornycroft used that ploy for a rise quite often! It worked for him; he ended up Chief Structural Engineer at BAE Systems Glasgow for the Type 45 Destroyer and Queen Elizabeth Carriers.

Two things came out of this move from Dowty-Rotol: 1. I probably jumped the gun too soon, and 2. If you go into an unsatisfactory job, get out as soon as you can; which I did. Don't soldier on thinking things will get better. Most likely they won't however much effort you put into trying to turn it round. Get out before someone gets their claws into you. That's really good advice; if you made a wrong move; get out fast. More on that later.

My digs in Bath were close to Sally's college, so I saw Sally weekends and during the week. I'd drive her home to Harrow Weald at end of term, and more or less be a permanent fixture at her home weekends until it was time to bring her back for the new College term. Her parents didn't seem to mind, made me most welcome; Sally's friend was their friend.

Driving down to Harrow Weald from Bath one Friday afternoon I was pulled over by a Police Car. I was last in a stream of traffic with this Police Car close behind me; I waved it past and in passing indicated I should pull in. The Policeman had a Cadet with him. The policeman said my rear light wasn't working. I said which one, he said onside to which I brightly replied that's not too bad then. The retort came back any more of that and I'll take your car in for a test and throw the book at it! Ain't some of our policemen wonderful! Next, I was asked to put on the handbrake and the coppers tried to push the car along; couldn't do that so handbrake OK. Cadet see's my exhaust pipe looked to be hanging low. Had to start engine, policeman puts hand over exhaust and declares silencer faulty. Got booked even though I'd said a replacement exhaust pipe was on order, couldn't buy it over the counter. Surprised I was booked because in those days 2 stroke Diesel Dodge lorries were common place on our roads; they made one heck of a noise; some had passed in the opposite direction.

When the summons came, I wrote to the court saying 'The case was waste of Police time, I already had an exhaust pipe on order. I thought the Policeman had either had a row with his wife or was showing off to the Cadet.. I asked why the Policman hadn't booked the noisy 2 stroke diesel lorries that had passed; they made a tremendous racket. I went on to suggest that had they booked one of these lorries he; would probably have found himself in the ditch one dark night'. I got fined of course, but the Bath Chronicle picked up on it under the headline 'Case waste of Police Time'.

Next day the paper published a picture of what appeared to be a perfectly healthy Swan being placed into the River Avon by a PC!

Perhaps fortunately soon after that I moved to a job at Bristol Siddeley Engines at Patchway and new digs!!

Policing is a challenging job. Policemen experience real life threatening situations in protecting the public, and indeed see horrific injuries in various incidents they attend. I'm sure we all admire what they do.

But I think both incidents I had with the Police, called for a Caution to get it fixed by a certain date, rather than a summons. In the first case, I was young and inexperienced; a word would have done the trick; Caution if you must.  In the second case, I already had the exhaust pipe issue in hand; that didn't need a summons or the heavy-handed bully boy one sided threats by the PC.

The Police need the publics good will; it is counter-productive of them to unnecessarily risk antagonising the public whose support is needed to pressure Government to adequately fund the Police Service.

However well the Police are thought of, they don't have the same place in our hearts as the NHS.  And I believe this is partly their fault; they need to be more people friendly. The Police image also needs dusting off; need more of the friendly neighbourhood copper; less of Cops and Robbers!

Having looked after myself from the age of 20, I tend to say what I think and don't take nonsense from anyone.  Perhaps not the ideal Civil Servant!

# 7. PROJECT SUPPORT ENGINEER @ BRISTOL SIDDELEY ENGINES LIMITED, PATCHWAY.

Bristol Siddeley Engines Limited (BSEL) was formed in 1959 by a merger of Bristol Aero Engines Limited and Armstrong Siddeley Motors Limited.

In 1961 the company was expanded by the purchase of the de Havilland Engine Company and the engine division of Blackburn Aircraft.

Bristol Siddeley was purchased by Rolls-Royce Ltd in 1986.

Bristol Aero Engines and Armstrong Siddeley Motors were the aero engine manufacturing companies of the Bristol Aeroplane Company and the Hawker Siddeley Group respectively.

At about the same time the Bristol Aeroplane Company was absorbed into the British Aircraft Corporation along with English Electric Aviation and Vickers Armstrong.

Bristol Aero Engines had been formed in 1920 when the Bristol Aircraft Company took over the asetts of the Cosmos Engineering Company who had gone into liquidation.

Cosmos had manufactured aero engines in a factory in Fishponds Bristol during World War 1.

In 1918 Cosmos built Rolls Royce Hawk and Falcon engines under licence.

In 1919 Cosmos had developed it's own range of radial engines: the 45ohp Jupiter, the 300hp Mercury and the 100hp Lucifer.

The company also produced a small number of cars. It got into financial difficulties, and was bought by Bristol Aero Engines.

After buying Bristol Siddeley engines, Rolls-Royce was one of the largest companies of its kind in the world at that time, offering a wider range of engines than any other manufacturer.

Aero engines produced by the company included piston engines, turbo-props, turboshaft, turbojets, turbofans, auxiliary power units (APU), ramjets and liquid propellant rocket motors.

The Bristol Engine Division (BED) of Rolls Royce; formerly BSEL; designed and produced the Orpheus turbojet for the Gnat trainer (also used in Donald Campbell's world water speed record breaking boat Bluebird, in which he was later killed in an accident on Lake Conniston whilst attempting to raise the speed record further). Cambell's Bluebird hit a swell on the lake at high speed; this caused the boat to pitch up, increasing the angle of attack to the airflow, and it took off somersaulting, and killing Cambell when it hit the water.

Other engines designed and produced by BED include:

The Olympus two spool turbojet for the Vulcan bomber, and developed in an enlarged form for the future V- bomber replacement; theTSR2 (cancelled). This engine was also intended for Concorde, which after the TSR2 cancellation then bore all the engine development costs, much of which would have been absorbed by TSR2. A significant reason why Concorde costs escalated so much; others being it was an ambitious cutting-edge project, years ahead of anything else other than the Russian Tupolov design.

The Pegasus two shaft medium by-pass vectored thrust turbofan for what was to become the Harrier VSTOL ground attack aircraft. The Harrier was operated by the RAF, Fleet Air Arm, the Spanish Navy and the US Marines, where for them it was built in the US under licence in a further developed variant; many of the features being incorporated in UK built Harriers for the RAF and Fleet Air Arm.

The Proteus turboprop for the Saro Princess Flying Boat, developed further for the Bristol Britannia airliner and in marinised form for hovercraft and warships.

The Viper turbojet used for military trainers and ground attack aircraft. It was also used in the Hawker Siddeley HS 125 business jet.

BSEL had another vectored thrust turbofan; the plenum chamber burning BS 100 intended for the Hawker Siddeley supersonic P1154 VSTOL fighter; the Harrier replacement. The P1154 was cancelled along with the BS 100 engine.

The M45 turbofan was developed with the French company SNECMA for the German VFW 614 feederliner. Although the aircraft flew, it wasn't proceeded with and further work by BED on the engine was stopped. I have a feeling SNECMA used the engine technology to develop an engine eventually for the Boeing 737 airliner, but not sure about this.

BED designed and produced the RB 199 turbofan for Tornado and the Hawk Trainer.

The company also were responsible for the RB 199 engine successor used in the Typhoon fighter.

The job with Bristol Siddeley Engines at Patchway was to support their Aircraft Project Analysis Group by preparing aircraft layout and engine installation concepts and making detailed weight estimates of these designs. An interesting job. Weight estimates were made according to a Royal Aircraft Establishment devised method. A lot of tabular work which needed detailed information regarding fuselage, wings, tailplane, fin and rudder, control surface areas, flaps and location and height of undercarriage etc. The effect of any unusual design features on weight had also to be assessed.

The Performace Engineers gave me the relevant basic data, wing planform, tail appendage, fuselage size etc. and I took it from there. If I could find a General Arrangement (GA) of a similar type of aircraft in Flight, Aviation Week and other aviation journals. I would use this as a starting point for the layout and adapt it to suit the given parameters.

I was tasked with developing a 'packaging routine' whereby based on a consistent method, civil and military aircraft GA's could be drawn out on a rational basis. This required perusing Aviation Magazines for aircraft GA's and drawing a number of plots: fuselage width v number of seats abreast, fuselage depth v width. Fuselages weren't all of strictly circular section.; many aircraft had a freight hold beneath the passenger cabin, giving the fuselage more depth than width. Other parameters were length and proportions of nose and tail-cone profile v fuselage depth, height of cockpit above bottom of fuselage and size of cockpit windows; height of tailcone above bottom of fuselage; tailplane area v wing area, distance between wing and tailplane centres of lift v fuselage length, galley area, toilet spaces, seat pitch for tourist, business and first class, width for boarding and emergency exits, external door sizes, typical fuselage frame spacing, and frame depth v fuselage width and so on. Developing a consistent, and rational technique for the layout of an aircraft fuselage.

I also defined standard cockpit sizes for fighter aircraft for single seat, two seat abreast and tandem cockpit arrangements.

So, with this information together with the wing size inputs from the performance engineers; using my draughtsman's eye I was able to draw a realistic looking aircraft. It was often said to me that now they could present a project that looked realistic and helped interest others in their presentations.

Through this work, I developed an eye for drawing up real looking aircraft, for which I was complemented on. But I couldn't see much future in it.

The role of the Aircraft Project Analysis Group was to assess existing aircraft projects and show the advantages of using a Bristol Siddley engine over the one currently chosen. Sometimes if a new aircraft design or stretch of an existing design was foreseen, the 'group' would work this up into a proposal and hawk it around the aircraft companies and airlines to try and gain interest.

Apart from the Viper turbojet, used in Military Trainers and the DH125 Business Jet; all Bristol-Siddeley engines had military applications, predominantly in the Vulcan bomber and Harrier vertical take-off and land (VTOL) fighter. They also had the new Olympus Engine being developed as a power-plant for TSR2, a project by now cancelled. This was virtually a new engine other than in name; it was larger and more powerful than the Olympus variant used in the V-bombers from which it was derived. The engine would now only be developed for the Concorde supersonic airliner. One reason why the development cost of Concorde escalated. Instead of TSR2 taking the brunt of engine development costs, this all fell onto Concorde!

The company were keen to find other civil aircraft applications for their engines other than Concorde; they needed to break into the civil market.

Rolls-Royce were developing the RB211; a new high thrust large diameter three spool by-pass engine for the Lockheed Tri-Star airliner. That engine, in modified form, would also be suitable for the proposed Airbus wide-body, medium range airliner.

Bristol-Siddley had nothing to offer in that size of engine. They explored developing the American, Pratt & Witney JT9D engine for an Airbus size application; Bristol Siddley were working on a new front Fan; the rest of the engine being built under licence by Bristol Siddeley. At the time the UK Government had a big stake in Airbus, so it was likely a UK engine; or one built under licence in the UK, would be the preferred choice rather than an imported US engine. The Bristol Siddeley engine development of the JT9D filled the bill!

In support of the companies interest in this engine; the Aircraft Project Group prepared a GA of an Airbus with these engines together with weight and performance estimates. And this was presented to Airbus showing the proposed Bristol/Pratt & Witney engine as an alternative power-plant to the Rolls-Royce RB211 engine.

France and Britain were developing the AFVG, a variable swing wing fighter bomber; the two companies involved were British Aircraft Corporation (BAC), English Electric Division and the French Aircraft Company Dassault. The preferred engine supplier was Bristol Siddeley. Eventually the French pulled out; maybe when they had found out how the variable geometry wing worked; and went off to build a VG Fighter Bomber on their own with a different engine and the agreement collapsed.

The British then teamed up with the Germans and Italians to develop their own Swing Wing Fighter Bomber; later named Tornado. Bristol Siddeley was to provide the new engine; later designated the RB199 after Rolls-Royce and Bristol combined. But at the early design stage of this engine the merger between Rolls-Royce and Bristol Siddeley was some way off.

Because of the strict security surrounding the Tornado project, English Electric Aviation; the Prime Contractor; were not allowed to give the engine designer Bristol Siddeley relevant aircraft drag and performance data so that engine design points could be established.

It was over then to the Aircraft Project Analysis Group to provide a GA of the Tornado based on published information, size the fuselage to include air intakes for the engine, weapons and fuel and to 'area rule' the aircraft for minimum supersonic drag. This involved much trial and error until the performance engineers were satisfied the best conceptual area ruling had been achieved. From this 'model', aircraft drag and engine design point could be calculated. This was presented to English Electric at Warton; and on this basis engine design parameters were agreed. Today this technique would be called 'reverse engineering'! I had no idea of the performance figures, my role was to prepare the GA and area rule the fuselage as best I could.

At about this time BAC (Filton); just up the road from Bristol Siddeley at Patchway started advertising a course for Engineers suitable for training as Aircraft Stress Engineers. I wrote off; got an interview; was accepted and asked if I could be seconded on the course as a Bristol Siddeley employee, something I had been promised at interview with them. My letter from BAC (Filton), offered me a place on the course, but said that would be conditional on my joining BAC. I took this letter to the Director of Projects at Bristol Siddeley, GT Smith, told him that at interview for my present job I had been told I would be placed on one of these stress courses at Filton when they came up again. And would it still be possible? I showed him the letter and said I didn't want to change companies. Mr Smith was the sort of man who supported his people. He arranged for me to be seconded on the course, but as a Bristol Siddley employee. I think my work must have been appreciated. The course was 3 months full time at Filton.

During this period I married Sally. When I returned from the course, I got a good rise, which mean't we could put a deposit down on a new small three bed detached house in Portishead, on the Bristol Channel. A comfortable commute over Bristol Suspension Bridge from work. Sally was working as an Infants Teacher at Pill Primary School; only a few miles from Portishead.

After I'd completed the stress course, I attended evening class at Bristol Technical College studying HNC Aerodynamics one evening a week. I thought this may broaden the scope of work I was given.

My thinking was that an Aerodynamics qualification combined with the aircraft structures course would enhance my usefulness within the Aircraft Projects Group and maybe lead to getting involved with performance estimates. A good thought, but it would have taken some years before I was trained up in that area. In any case who would do my job. I would never have been able to compete with the honours graduates doing the aircraft performance work

I thought quietly about where the expertise I was building up in the project support work could lead? The answer was nowhere! It was an interesting relatively well- paid job, but as time passed, I would loose my designers skills and usefulness elsewhere. If I was made redundant in the future my skill base would be specialised and very narrow. Something I had always tried to avoid; keep a broad skills and experience base was my mantra. And as described later, that paid off handsomely.

Stressing the engine installation appealed to me, but I soon realised that without a mentor, I was going to be up against it trying to develop conceptual layouts into provisional designs. It was highly specialised work. I didn't have the expertise; it wouldn't happen. Specific design work would go to the stress office if a customer expressed interest in a particular proposal by the group. My job was the preliminary conceptual and weight estimate, not stress analysis.

In the Project Analysis Group I'd got as far as I could. So, time to move on and consolidate. That's another useful tip: stop and think every now and then where the job is going to lead. You can't always avoid a dead end; most jobs have a glass or otherwise ceiling. But always keep your skill base intact and so keep yourself employable with a range of skills and experience to offer. And keep the skills current.

# 8. STRESS ENGINEER – POWERPLANT INSTALLATION @ BRISTOL SIDDELEY ENGINES, LATER ROLLS ROYCE (1971) LTD, PATCHWAY.

I asked for a transfer from the Aircraft Project Analysis Group into the Bristol Siddeley Engine Stress Office. People tried to persuade me out of it; but I said I couldn't see any future where I was and wanted to get involved in powerplant installation stress analysis; something tangible.

I was accepted by the Stress Office on a 3- month trial basis and put in Ian Laynes Section. He was responsible for stressing jigs, engine packaging to protect the engine in transit and handling, and manufacturing concessions on engine parts.

Once a jet engine being transported fell off a lorry; the driver phoned in to say one of your cement mixers has just fallen off my lorry! To the uninitiated it was a round object and could have been taken for that.

My first job was to stress a large jig for manufacturing an engine cowl for the Rolls-Royce RB211 engine which was to be flight tested on a VC10 Airliner Flying Test Bed.

After this trial period, I was moved to the Power-plant Installation Section working under Bernard Grant (previously mentioned, who I noticed walking down the back of the DO at Warton with the Bristol Siddeley contingent for the Olympus/TSR2 project) and Roy Williams. Both highly intelligent, open minded kind and clever men who took me under their wing and had the patience to train me up. The Group Leader was Wolf Gittleman. Another very clever decent man; I was lucky.

My first job was to recalculate the Production Concorde Engine Mounting Analysis following the Proto-type calculations as a guide.

Then working on the Olympus 593 engine itself; the ant-icing, gearbox mounting, fuel and oil pipe runs. Engine carcus dressing; these items were termed. Allowance had to be made for flight loads and the relative thermal expansions between the engine casing and the ancilliary's and pipework.

The gearbox had a hard mounting at the power take off shaft, but knife thin supports that allowed bending in one plane, but rigidly supported in the other. These knife supports were arranged at 90 degrees to one another and were at the extremity of the gearbox casing, thus providing gearbox support for flight loads and allow differential thermal expansion between the gearbox and engine casing. Another job was to stress the Oil Tank attachment to the engine casing and the anti-icing system where similar considerations had to be taken into account. Fuel and lubricating pipe-work had also to be stressed. These were hard steel pipes.

Power-plant stress analysis was very different from that of airframe stressing. I did a little aircraft stressing on the VC10 RB211 Flying Test Bed conversion work, but not in sufficient depth or long enough to become proficient in aircraft structural analysis.

It takes a number of years to become a proficient Aircraft Stressman. That would have been my ultimate aim given the choice, but my career took a different path. Maybe if the firm had got a new engine order and more Flying Test Bed conversion work I would have progressed in that area, and I would have stayed, but the Industry was beginning to run down; and New Zealand beckoned; but that was later.

Maybe it doesn't seem there could be much stressing involved in pipe runs, but you would be wrong. The pipe geometry and section constants would be input into a computer program. In those days based on beam elements and written by Roy. One end of the pipe was assumed fixed; the other released, but loaded by displacements. These were the differential thermal displacements between engine casing and pipe for various design cases: cold start, where say the aircraft had been cold soaked in a Canadian winter overnight; on start up the engine casing gets hot faster than the ancilliaries, which for this case are pessimistically assumed cold; the in flight cruise, and hot start cases would be others.

For these conditions pipe bending stresses are calculated and combined with pipe hoop and longitudinal stress. Pipe attachment stresses; flanging and bolting to the engine casing are also checked. Allowable stresses must not exceed Creep Stress at a specified design case high temperature; an important criteria. Creep stress is the stress at a high temperature where the metal will continue to permanently stretch without specified limit. It is lower than the Proof Stress at room temperature. Proof Stress is the stress at which a specified permanent strain occurs. Ultimate Stress was the stress at failure. Neither, Creep, Proof or Ultimate stresses must be exceeded.

The Proof Stress limit would relate to a non high temperature design case; say during maintenance where a pipe flange may need to be disconnected to replace a seal and a certain amount of pipe displacement may be required to get at and replace that seal. It is the amount of displacement that can be allowed without exceeding Proof Stress and permanently straining the pipe that has to be calculated.

Another consideration would be the effect of a worse combination of manufacturing tolerances; 'lack of fit' in stressmans parlance, where the pipe would need to be pulled up by the bolts to fit. These built in stresses would also need to be considered in the stress analysis, although in practice the effect was small and could usually be discounted. In flight 'g' loads were usually insignificant when pipe stressing, unless the pipe supported an unsupported weight, in say a 'fail safe' case.

The program would be re-run with the displaced end fixed and the former fixed end freed, but now 'loaded' with the previous displacements to find the flange and bolt loads on the other end of the pipe. These stainless steel pipes were known as , 'hard pipes' and were used because they were more dependable in service than the braided proprietary flexi-pipes.

Loadings had also to be applied to the engine casing flanges from these ancilliary mountings. The flanges are analysed as rings to RAeS Data Sheets, whereby stresses and displacements can be calculated.

I carried out similar work on the M45 Turbofan engine; designed and developed with the French company SNECMA. This engine was intended for a small German feederliner of about 40 seats; the VFW 614. This aircraft was unusual in that the engines were mounted on above wing pylons rather than the conventional below wing arrangement. The reason for the over wing engine mounting was to minimise debris ingestion by the engine on unmade runways. This concept however was found to have a serious flaw; the wing would oscillate in flight due to turbulence, and then set the engine pylon oscillating. When the wing oscillations reduced, the oscillating engine pylons re-excited the wing. The problem was overcome by Roy Williams who designed a torque tube to dynamically decouple the engine mounting from the wing. Problem solved. I think two aircraft were flown; one was lost due to flutter. The project was soon cancelled after that.

It was during my time in the stress office that I applied for and obtained Membership of The Royal Aeronautical Society and gained Chartered Engineer status.

Midway of my time with the power-plant section, Bristol Siddeley Engines merged with Rolls-Royce. It was popularly described as a marriage; Rolls-Royce had virtually all the civil work, Bristol-Siddeley all the new military engines; the JT9D proposal had been dropped earlier. By default Bristol became centre for Military Engines, and became the Bristol Engine Division (BED) of Rolls-Royce.

Dr Stanley Hooker is famously reported as saying 'in a marriage someone always gets f…..d'

Apart from the name change and transfer to the Rolls-Royce Pension Scheme, everything remained the same at our level.

Rolls-Royce at Derby were developing an all new three spool bypass engine; the RB211 for the Lockheed Tristar wide body airliner. The engine in developed form was also a contender for the new Airbus and Boeing 747 airliners.

The original RB211 design used a carbon fibre fan. The carbon fibre fan had been tested and a carbon fibre fan fitted to a Conway engine on a BOAC VC 10 airliner to test a carbon fibre blade in service. This worked well, but when a full sized carbon fibre fan was flight tested on the RB211 engine, there were failures. The fan design had to be changed to one made from Titanium. Other problems were found with the engine causing delays and forcing Rolls-Royce into bankruptcy. Dr Stanley Hooker was called in to sort the engine out. It was very serious for Lockheed as well who had designed their Tristar airliner around that engine. It's probably true to say potential Tristar orders were lost to the DC10 airliner; an aircraft not without serious shortcomings shown up later in service.

Rolls-Royce wanted a step change in the design of this new engine that put it ahead of the American engine companies Pratt & Witney and General Electric. To do this Rolls-Royce got together their brightest graduates; the older experienced men had the rol'e of advisors to filter out hair brained schemes and eliminate known design faults from being repeated by inexperienced graduates. It didn't work like that. The Graduates by passed the experienced designers in their quest for the 'step change'. Design development was essentially uncontrolled; ending in tears and bankrupting the company.

After bankruptcy the company re-emerged as Rolls-Royce (1971) Ltd. The Car Division had been sold off, but retained the Rolls-Royce name.

Management consultants were crawling all over Derby. There were big redundancies. At Bristol, we were awaiting our turn. We were left alone. It was the Bristol Engine Division making the money, but for a while there was big uncertainty.

Jobs were being advertised on the Zambian Copper Belt. Three year contracts were offered with free housing and medical care and schooling. The money was good and it would have been possible to buy a Ford Cortina, take it out with you, rent out our house in Portishead, and come back with the car after three years and pay off the mortgage! Well it seemed it might be possible. I wrote off, got an interview in London, was offered a job, but not at the salary I'd hoped for. Not surprising really because I had no heavy Engineering or Mining experience. Some of the people I met going for interview were old hands. They said the housing was fine, but you needed a high fence and Alsation to keep the locals out; there was a lot of thieving; they had very little. That was something of a shock. Zambia didn't seem such a good idea anymore!

However as my job now seemed safe at Bristol, I did no more about it. But it set me thinking about emigrating to New Zealand; my old Grannies words when I was a boy had imprinted in my mind; more of that later.

Few Concordes were built. The project was slowed by the Ted Heath Conservative Government to save cash flow on the project. America and the Environmental Lobby prevented the airliner flying supersonically over highly populated land mass, and tried to ban it from operating out of La Gaurda Airport New York because of noise during take-off. Tests showed the aircraft was within the new legislative limits so it was able to enter service. But the big jump in fuel prices also had a serious effect on operating costs.

The delay into service rather played into Boeings hands. They were bringing their new 747 Jumbo jet into service at about the same time Concorde was originally scheduled for introduction to service. Own goal by HMG?

The Boeing 707 was at the end of its stretch and development potential. Boeing needed a replacement. The B747 had been configured originally for a military freighter project; hence the raised cockpit to allow nose doors and unobstructed access to the hold.  Boeing lost out in the military freighter competition to Lockheed's Galaxy; an enormous military freighter aircraft, much larger than the B747.

Boeing were able to reconfigure the military freighter concept into the 747 high capacity long range airliner on the back of the US funded 747 military freighter project.  It could carry about 3 times as many passengers as a 707, with lower seat mile costs.  It was ideal for the high density trans-Atlantic service.  The new Boeing airliner would also reduce airport landing congestion which was becoming an issue.  Maybe the airline flying the big aircraft retained its number of landing and take off slots.  If so it had increased its potential passenger capacity at the 'gate'.

The 747 had greater range than Concorde.  So, the 747 could overfly Concorde's refuelling stops on extra long distances, so reducing the Concorde's speed advantage.

Concorde had been designed to fly from London or Paris to New York; the limit of its range.  Supersonic flight restrictions over land would have been a serious blow to an airline expecting to use the aircraft on trans-continental flights, it wouldn't be allowed by new legislation.  The Concorde had been relegated in effect to overwater routes, and so damping interest from the airlines.  Without a strong sales base for the first version, there wasn't a demand for developing a larger longer range aircraft.  And so the project drew to a close.

Only twelve Concorde's were sold; six to British Airways and six to Air France at knock down prices.  However these airlines operated the type for many years and made big profits from the operation.

With few Concorde's in service, few Olympus 593 engines were needed.  Further engine developments would be very limited.  Anything major would be cost prohibitive.

When design work on the Rolls Bristol, RB199 engine for Tornado and Hawk was completed, the company needed a new engine on its books.

The Eurofighter project would have only been in embryonic stage at that time; the engine was a long way off before we would get involved. If indeed Rolls Royce Bristol would develop it; the contract so far as I know had still to be finalised.

We were still busy, tying up loose ends. But we would need something new before too long I reckoned; my previous experiences getting my antenna working!

Salaries were relatively low, but so far as I was concerned, I had a very interesting job, was being trained up in a good friendly and highly professional work environment.

I wasn't likely to get a promotion; but that didn't bother me; the job was great. We would get cost of living rises, plus if you were lucky a merit increase. The merits were: £25, £50 or exceptionally £75 per annum. You may get a merit every three years or so, but there was no guarantee, there wasn't a lot of money for the company to spread around. I regularly did Saturday morning overtime once a fortnight to pay for extras.

We used slide rules for calculations unless great accuracy was required. Then we would use an electronic calculator more or less half the size of a small desk! The firm brought in a new high tech. small hand held Scientific Electronic Calculator they screwed into a block of wood so it couldn't get lost! The firm offered us a deal: if we bought one of these calculators through the firm, they would supply batteries and service it for free; the cost to us individually was £75 for a calculator! In those days £75 was a lot of money; my mortgage was £36 a month and there was little left over from my salary at the end of the month. We couldn't afford it. I didn't know anyone who took the offer up.

The Directors probably thought we were a mingy lot! But with a young family there were more pressing needs. We didn't know when or what the next merit rise would be; getting a merit, and over the £25 minimum and you would be very fortunate indeed. Cost of living rises were absorbed; they never really fully compensated for inflation, we were as strapped for cash as we were led to believe the firm was. Directors didn't get the obscene rises and bonuses of today; the Unions saw to that!! The money had to be seen to be fairly spread around. And I think the attitude of management was better for it.

I was a member of the Engineering Union. Around that time I drew the short straw at the Office AGM. It was my turn to be Secretary of the Office Committee. Previously we had voted on putting together a wage claim for the office. I thought; what have I got myself into here? I did a Stress Office survey inviting people to mark their annual salary against age on a graph. A few spurious plots were put on the graph to encourage people and for privacy; I looked away whilst this was done.

The resulting office salary plot was interesting. It showed a bow wave effect; the younger members with families and mortgages were often doing as well or better than older more experienced men; the backbone of the stress office.

Remuneration to a large extent seemed to be based on perceived need and the fact older men who had paid off the mortgage and whose children had left home were probably managing OK. And with long service and pension to look forward to they didn't want to move. On the other hand when somebody reached 60, the firm usually gave them a good rise to help boost their pension. But that wasn't a given; management practice could change; and there would be a certain amount of luck. You had to hope you weren't out of favour when your turn came around!

I presented this salary plot together with a number of advertisements for similar calibre of jobs in the general area together with covering notes in a report which was the basis of the claim. The Office claim sub-committee had a meeting with the Chief Stressman and his deputies. I ran through the claim and handed a copy across the table.

I held a Stress Office meeting one lunch hour when union members and a number of non- union members turned up. Senior well respected stressmen turned up for the meeting; so we'd made our point. Pertinent details of the claim were shown to them and they were updated as to where we were.

For most of them 'Trade Unions weren't for them', but they were impressed the way the claim had been presented. That they turned up at the meeting was a good sign of their interest, and must have sent a strong message to Management.

The managers were sympathetic to our claim; they appeared to welcome it, because it did give them some ammunition. They were aware salaries were low and falling behind other areas in the company with strong union representation. But I think that genuinely the coffers were low, and work was running out.

I put a lot of work into the claim, at home and I was allowed to do what I had to at work. It was a bit depressing really because although it might rattle a few chains, I doubted it would make a big difference. It was passed up to Director level I was told unofficially.

The claim brought me to managements notice; they gave me more interesting and absorbing work, modelling the Olympus engine casing, and applying flight loads from the mountings and ancilliaries to check casing displacement to ensure turbine blade tip clearance was maintained. Other interesting work followed; the wage claim; and how it was managed; did me some good. It could have gone the other way of course, it was a risk I took, but that was a measure of the company management. Decent and fair minded.

I was getting a little concerned that we needed work from a new engine, but nothing seemed near, no new military projects apart from Eurofighter in the distance.

About this time New Zealand Electricity were advertising for Design Engineers for their new Huntly Power Station. I applied, but not having related experience, did not get an interview. My grandmother had once said to me when I was about 11 'you should write to Alice (her daughter in New Zealand), one day you may want to go there'! Sometimes things stick in your mind if you are ready to hear.

At the same time I was preparing the claim, I researched the New Zealand option and found out what I could. I wrote to my Aunt Alice, who lived in Auckland, sent off for NZ newspapers, found another contact in New Zealand who was happy to write to me. He'd emigrated many years previously; had gone out as a draughtsman and was then Manager of the Dulux Paint Works in Lower Hutt outside of Wellington. He was enthusiastic about NZ and the lifestyle. His name was Ted Lissamen, he lived in Stokes Valley, near Wellington. He was near retirement. By the time we started emigrating, he'd gone to live in Whangeri, near the Bay of Islands; nice spot. We met up later.

I wrote off again to New Zealand Electricity in Wellington, and was invited for interview at the New Zealand High Commission, at New Zealand House in the Haymarket, London. Moral; if at first you don't succeed, try again; it worked!

# 9. EARTHQUAKE ENGINEER @ NEW ZEALAND ELECTRICITY, WELLINGTON, NZ.

Initial use of electricity in NZ was for mining operations. The first industrial hydro-electricity generating station was established at Bullendale in Otago, South Island in 1885 to provide power for Pheonix mine.

The plant used water from nearby Skippers Creek, a tributary of the Shotover River.

Reefton on the West Coast became the first electrified city in 1888 after the Reefton Hydro Power Station was commissioned.

The first sizeable power station was built for the Waihi gold mines at Horahors on the Waikato River.

This set a precedence with hydro power becoming and remaining the dominant source of electricity. In 1930, hydro generated 92% of electricity generated.

Industrial usage quickly increased, but it was government programs in the first two thirds of the 20$^{th}$ century that caused private demand to grow strongly, to the point that from 1936 on, power shortages began to occur. The large number of power stations built in the 1950's enabled supply to catch up again.

All the government energy assets originally came under the Public Works Department.

From 1946 the management of generation and transmission came under the newly formed State Hydro-Electric Department (SHD). In 1958 it was renamed as the New Zealand Electricity Department (NZED).

In 1978 the Electricity Division of the Ministry of Energy assumed responsibility, renaming it New Zealand Electricity. Distribution and retailing was the responsibility of local electric power boards (EPBs) or municipal electricity departments (MEDs).

New Zealand electrical energy generation, previously state owned, as in most countries, was corporatized, deregulated and partly sold off over the past two decades. Much of the generation and retail sector remains under government ownership as state owned enterprises.

In 1987 the 4$^{th}$ Labour Government corporatized New Zealand Electricity as a State Owned Enterprise, named the Electricity Corporation of New Zealand (ECNZ) trading for a time as Electricorp. The Energy Companies Act of 1992 brought in by Labour, went further, requiring EBPs and MEDs to become commercial companies in charge of distribution and retailing. Thus began a feeding frenzy; the fat getting fatter, whilst many NZE employees were made redundant and their lives ruined. By coming back to UK in 1980 I escaped all this.

In 1994 ECNZ's transmission business was split off as Transpower.

In 1996 ECNZ was split again with the major assets formed into three new state owned enterprises: Mighty River Power, Genesis Electricity and Meridian Energy. Minor assets were sold off. At the same time local power companies were required to separate distribution and retailing, with the retail side of the business sold off mainly to generation companies.

As a country New Zealand still suffers from the EU ban from exporting to Europe and UK. A great thank you from the UK Heath Conservative government to those thousands of New Zealanders who lost their lives or were maimed fighting to defend the UK and Commonwealth!

I went with my wife Sally to the interview at New Zealand House. I got the offer of a job at New Zealand Electricity Department's Development Division in the Research Section investigating seismic response of power station equipment and distributive switchyards. R&D work was also involved. The salary was excellent and progressive year on year. I took up the offer. It was a NZ Government Department job, so should have been for life.

I carried on with my work at Rolls Royce. I really didn't want to leave, but the offer was too good to refuse. It took about 18 months from job offer to get the visa and sell the house. We took Sally's widowed mum with us as well.

We went out by sea, with the Greek Lines Xandros on the Australis ex US Lines America, a 45,000 tonner. A beautiful ship, good cabin. We called in at the Canaries, Cape Town, across a stormy Indian Ocean to Sydney, where we were delayed 2 days because the ship failed a safety inspection; the lifeboats couldn't be lowered! Sally's Uncle Bob and his wife drove down from Brisbane to see us. Bob was Sally's fathers brother, they hadn't seen one another for about 20 years. It was then Melbourne and across The Australian Bight to Freemantle and a taxi tour of Perth. The final sector was across the Tasman Sea, through the Bay of Islands and finally Auckland. We were warned on the ship there was going to be a crew go-slow so I had to get all our baggage from the Luggage Master and into our cabin with the cabin baggage and take it off at Auckland. Otherwise we were told it would stay on the ship! I scouted around the ship, found a Porters trolley, hid it in the cabin; loaded it up and wheeled it off the ship at Auckland. The last I saw of the trolley was a Kiwi wheeling it off the dock! We were at last in New Zealand. A fantastic journey.

We were met by our 'Pen Pal', Ted Lissaman and his wife. He'd retired and moved up to Whangerai; North of Auckland; but had driven down to Auckland to meet us off the boat. Very nice of him and his wife to come and greet us. Later when we drove up to the Bay of Islands on holiday we called in to see them.

I'd hoped my NZ cousin Leo would be there to meet us. He wasn't. Our pen friends the Lissaman's drove us to Aunt Alice's bungalow. She was in hospital; which is where Leo probably was. We didn't know she was seriously ill; so, left it at that; thinking we'll be up here again someday and see her then. But I heard later she'd died. Just missed her, my father's sister.

New Zealand Electricity Department (NZED) had originally booked us on the Silver Fern; a diesel railcar that took you through New Zealand down to Wellington. It would have been a fabulous journey. But because of the ship's delay in Sydney, when we reached Auckland, the train was full, so we flew down in an Air New Zealand Boeing 737. Boy, was it windy at Wellington Airport! On landing the aircraft tipped up almost onto its wingtip; a hairy moment! Windy Wellington; we had arrived!

We were met by an NZED representative at Wellington Airport, taken first to a supermarket where we got $50 of groceries and then to our Motel.

I was picked up next morning and taken into work. Shared Office with another Engineer; Tony Rutledge; carpeted floor, own telephone; quite palatial to anything I'd seen previously. I was on the $7^{th}$ floor of a new 14 floor building, designed to withstand earthquake; good! In 1858 the ground it was standing on had been uplifted from the sea by a major earthquake; a 400 year event apparently; though soon after we first arrived the local paper was trumpeting about a significant earthquake being overdue. Not good news. Thirty-five years later it's still overdue; so when it comes it could be a big one! There was recently a large earthquake in Christchurch in South Island; not a high- risk area! Perhaps that has left Wellington off the hook for a while longer; but it will come eventually.

The shore on the Wellington side of the Cook Straight is being thrust upwards. On the opposite of the Cook Straight; South Island; the ground is slumping. In millions of years time; North and South Island will form an L, and even later, NZ will have been moved to where Hong Kong is now. That was good news to Stan Wong, who later became our Chief Development Engineer! He was positively looking forward to it.

I began work as a Senior Engineer in the Research Section. This had been initially set up in the Development Division to allow Harry Hitchcock full reign. He was an energetic dynamic personality; who among other things had developed a strong interest in seismic withstand of Electrical Equipment in sub-stations, Power Stations and so on. He had published a couple of papers on the topic in the New Zealand Earthquake Engineering Society Bulletin. He wrote many of the Earthquake Engineering clauses for Huntly Power Station which was still being built when I arrived. Tony Rutledge was the other Senior Engineer in the office working on earthquake withstand; a super intelligent laid back Kiwi, great to work with.

Harry's other interests were bio-mass and wave energy. He was coming up to retirement in a couple of years and was looking for somebody with an Honours degree or PhD to take over from him and the interests he'd developed in the Research Section.

What eventually happened was that Dr. Hank Bauer took over the Research Section as Senior Research Engineer when Harry retired and set up an Earthquake Engineering and Energy Group's. He brought in two Engineers to man Energy. I was made up to Earthquake Engineer. But first I was made Acting Earthquake Engineer for 6 months; then the job was advertised in the NZ Civil Service Circular. I applied for and got the job in open competition; of which I'm extremely proud.

But going back to the first days: I was taken along and introduced to the Earthquake Engineering Committee under the Chairmanship of Ian Crabtree a Principal Engineer from the Supply Division in Head Office. His interest in Earthquake was to keep Electrical Distribution functioning through and in the aftermath of an Earthquake. One way of doing this was to make Engineers in the 8 District Offices Earthquake aware. To do this it was decided to set up Seminars at Head Office. I was tasked with co-ordinating and organising this and the speakers.

Later when promoted to the newly created position of Earthquake Engineer I would take this concept on the road with Tony and Ian a graduate who was also working with me at the time. I would phone up a District Office; get them to invite us down; we would fly down and inspect a sub-station, take photos advise on how to improve seismic withstand, which after you had carried out analyses on similar equipment, a strengthening remedy was in many cases obvious and straightforward. Spend the night in a Hotel. Give the Seminar next day and fly back to Wellington.

NZED had eight District Offices covering North and South Island. All the Hydro and Steam Power Stations had accommodation available most times where NZED staff could stay with their families and eat in the canteen. More like a Restaurant; custard and ice-cream!

When you wanted to go on holiday, you phoned up the District Office, found out where there were vacancies and booked in. It cost $1 per person per night for accommodation and $0.5 per person per meal! Remarkable value, and excellent bungalow or motel standard accommodation. We were able to tour North and South Island this way at little cost.

But I'm going ahead of myself.

When I first arrived in the office I was put to work reading relevant earthquake engineering papers and then analysing a reinforced concrete turbo-generator pedestal for the new Marsden B steam power station in North Island near Whangerei. The analysis was straightforward using longhand; estimate the pedestal stiffness and calculate its predominant natural frequency. From the New Zealand Electricity Earthquake Response Spectra plot, and assuming a certain percentage of structural damping (3% I recall) read off the acceleration felt by the turbo machinery in a design level earthquake. The Contractor then had to ensure the thrust bearings and bolting of the turbo-generator to the pedestal could take these loads without damage. The bolting attachments needed strengthening as a result of this investigation.

Other early work was to write notes for and co-ordinate the lectures and presentations for the in-house Earthquake Engineering Courses to be held at Head Office in Wellington and liase with the 8 District Offices who were sending staff. About this time NZED had a name change to New Zealand Electricity (NZE); nothing else changed in the way we were organised or worked.

Eventually Harry Hitchcock retired and was taken on by the Supply Division working as a consultant to Ian Crabtree investigating seismic withstand and strengthening existing equipment throughout the New Zealand Electricity system.

Meanwhile we in the Research Section had a new boss; Dr. Hank Bauer; an American; retired US Army Colonel and former lecturer at an American New England University. A very decent man, but boy was he bright; when he switched on you could feel the waves! He had previously held a post in the Development Division as a consultant on new energy systems.

The Development Division was an interesting multi-national division to work in: Maori, New Zealanders, Chinese (later we had a Chinese Chief Engineer Stan Wong), Korean, someone from the Phillipines, a Canadian. Adam Snarski; ex Polish Army joined us for Smoko (Tea Break).

Hank set up two small Sections; one to investigate Bio-mass, Wind and Wave Energy, and the other to continue with Earthquake Engineering predominantly for new build and carry on with inspections and spreading the message through seminars held in the District Offices. .

After Hank took over, I together with Tony wrote an Earthquake Engineering Manual; a three volume tombe, including theory in a digestible format with many typical examples of current NZE equipment. We put everything into it to get the information out to the Divisions at Head Office and the District Offices.

I also found out how to use the Ministry of Works and Development (MWD) Strudl-Dynal finite element structural dynamic program from Roger Blakeley, the NZE General Manager's son. I used this program to analyse the seismic response of the new Huntly Power

Stations steel Turbo-generator pedestal. Tony Rutledge used the program to analyse the seismic response of main steam pipe; incorporating Shok-Lock dampers. These had a recirculating ball mechanism that allowed slow movements caused by thermal expansion of the pipe; but immediately a shock was felt; the damper temporarily locked solid. The placement of these Shok-Lock dampers was designed to limit the natural vibrational frequency of the steam pipe to that which would not be excited by earthquake ground shaking.

I showed the Supply Division how to use the program so they could use it to analyse their in-service equipment; I was doing my job of passing on new techniques to interested groups.

Transformer foundations were an area of interest and the response of a transformer to earthquake motion. We got the MWD to set up a rotating vibrator on an out of service Transformer in a nearby switchyard; got it instrumented and vibrated it till it was rocking! Interesting response; the porclain insulators were mounted directly to thick steel plate which was a relatively flexible mounting with the porclain tips being greatly displaced. Only way to overcome that was to have a lot of slack in the dropper cables to prevent tightening and breaking of the brittle porcelain. As a result of this work standard installation procedures were improved and a monolithic foundation technique was put forward to reduce seismic response by incorporating soil damping effects with these large foundations which could be used to mount transformer banks. Soil damping of building foundations was being investigated at Auckland University; we plugged into that idea for transformer foundations and got the approval of Auckland University.

I was asked to take on NZE secretaryship of an inter Government Departmental Committee looking into Geological and Safety Aspects of Nuclear Power in New Zealand. I joined after the major work had been completed. I took notes, issued minutes and arranged meetings. A very interesting time none the less; getting to know experts in their fields from the Ministry of Works and Development

(MWD). Geological Survey (GS) and the Division of Scientific and Industrial Research (DSIR).

The New Zealand Standards Association set up a working group to develop a new Code of Practice for the Seismic Restraint of Building Services. I represented NZE in this working group, making further contacts and cementing old. Dr Ivan Skinner represented the DSIR. I got to know Graeme UpRichard an MWD Engineer from Christchurch Division who dealt with Seismic Resistance of Building Services and had written a paper on the topic, which became the starting point for the NZ Standard.

The need for a Building Services Code of Practice was because the NZ Building Code of the time simply specified a side loading to represent earthquake effect, which in itself was much lower than the dynamic loading effect of the earthquake. Vertical accelerations were not taken into account, neither the effect of structural amplification on Building Services, ceilings etc. The Code was simply to provide earthquake withstand for the main structure.

To be effective this earthquake side load provision in the Code, had to be combined with a building structural ductile design philosophy. In simple terms columns needed to remain essentially elastic, but beams forced to absorb energy by plastic hinges adjacent to the column attachment. This needed beams with sections capable of forming 'plastic hinges' and good attachment detail design that would transfer these loads into forming the 'plastic hinge'. What about reinforced concrete? This had sufficient steel reinforcing, both longitudinal and transverse to prevent the loss of concrete section other than cracking, a feature that caused damping and absorbed energy. Around the 'plastic hinge' region extra wrap around steel reinforcing was used, as they said, 'sufficient to stop a canary escaping'!

Canterbury University at Christchurch and the MWD at Wellington did major research on Earthquake Resistant Reinforced Concrete Structures.

At the time of the early building codes, Architects took the relatively low side load as the earthquake load that had to be applied to ceilings and building services with no allowance for amplification effects. With later knowledge this was proved to be largely ineffective, unless combined with revised design techniques.

The new Building Services Code of Practice provided these design techniques. It gave realistic loads that Building Services would be subjected to in Earthquake and many other simple low cost techniques that would allow Building Services to survive and function after earthquake. Important for minimising damage, personal survivability and getting life back to some normality as soon as possible after the event. Provided of course the Earthaquake wasn't of a catastrophic level.

New Zealand Electricity's Earthquake Engineering Committee was the focal point for earthquake related equipment failures which were reported to the Supply Division and then filtered down to the 'Committee'.

One recurring problem was the tripping of Bulchov Relays in transformers. The relays were either of 'Reed Switch' operated or used 'Mercury Switches'. The relays sensed vibration within the transformer. Over a certain intensity these relays shut the power down. This was to prevent short circuiting should a transformer malfunction or an unknown external event cause mechanical shock and vibration. As a precautionary safety measure the power was cut off.

Earthquake tremors, felt regularly within New Zealand and even heavy trucks would cause the more sensitive Bulchov Relays to trigger electricity shutdown. The Mercury Switch Relays were usually very sensitive and so as far as possible used in the least seismically active regions of NZ. Seismic Resistant Relays were marketed by some manufactures, but they were expensive. I collected a bunch of various types of relay including the seismic resistant relays; took them to the MWD and got them to set them up on a shaking table and subject each one to a series of Earthquake motions.

The result was most interesting; some standard relays out performed the so called 'seismic resistant' relays which had a price premium for that facility!

The route to that of General Managers was by a system of musical chairs through the positions of the Chief Divisional Engineers. When I was there, a preferred route was: Chief Engineer (Development), Chief Engineer(Design & Construction), Assistant General Manager, and then the top job, General Manager. The Chief Engineer Development seemed to be an important position to have held in the race to the top. Development was a small Division in the scheme of things sited only in Head Office, whereas the other Divisions were much larger and represented in the District Offices which were organised along Head Office lines. That was my impression.

In the aftermath of Ted Heath taking the UK into the EU; meat, fruit and dairy produce from New Zealand (NZ) and Australia was severely curtailed in favour of EU agricultural produce. Australia had huge Ore deposits in Western Australia which they mined and exported. Today China is a major customer. These Ore exports keep the Australian economy vibrant.

NZ relied on sheep production (meat and wool), dairy products (cheese and butter) and fruit exports to maintain its economy. The UK was a major importer of these products before membership of the EU. Some thanks from Britian for the sacrifices of New Zealanders who fought for us in two World Wars to defend the Empire.

The NZ Government tried to fill the gap in their export earnings by building more Power Stations, both Hydro and Sintered coal/Gas fired steam power stations (Huntly) to provide cheap plentiful power for a new White Goods Industry initiative.

NZ was going to produce and export Fridges, Freezers, Washing Machines, Television sets etc. in a big way; out Taiwan, Taiwan! It didn't happen they didn't have the infrastructure to do this.

More energy for Industrialisation seemed to be the answer to the Governments prayers. Long term, they were looking very seriously at Nuclear Energy.

It is helpful to understand this diversion because it had a significant impact upon the eventual selection of the NZE General Manager.

Our current Development Division Chief Engineer had gone on to be Chief Engineer Design and Construction Division; a heartbeat away from that of Assistant General Manager and onto the top job.

The incoming Development Division General Manager Kevin McCool; was a former Avenger torpedo bomber Navigator in WW2. Looking at latest power demand forecasts he realised that NZ with the current power station building program would be able to generate far more power than it needed for years to come. Industrialisation hadn't happened nor was likely in the foreseeable future on the scale envisaged.

When the Assistant General Manager retired, that post was initially filled by the Design and Construction Chief Engineer on the traditional new power station building program ethic. He either hadn't realised NZ had more generating capacity than it needed, or perhaps he didn't during the selection process want to be controversial! I don't know, but I'm sure he would have been able to read the latest forecasts.

Kevin McCool challenged this appointment based on a policy of a continuing power station building program. His challenge was that in future NZE would have to face de-commissioning its older stations rather than build new because the generating capacity was already more than needed for the foreseeable future. In Government Service, Civil Servants are allowed to challenge an appointment if they feel they have a case. Kevin McCool did just that and won the challenge. One wag (Graeme Alderton) had stuck a plastic sword hilt to a large commemorative stone just outside Rutherford House, with the placard 'whoever pulls the sword from the stone this afternoon will be crowned king of New Zealand Electricity.' Kiwi slang for a joke is 'a dag'; don't know why; just a bit of extra information.

Kevin McCool went on to become General Manager. The Design and Construction Chief Engineer became Assistant General Manager.

He wouldn't have realised it at the time, few would, but Kevin had been handed the poisoned chalice, because in later years he saw the Privatisation of NZE. Nothing he could have done about it. It was Government policy.

Before I left, the NZ Government set up a Department of Energy (DoE). This took over forecasting energy needs and new energy systems from New Zealand Electricity. Clearly something was afoot. But it was only years later this became evident when NZE was privatised. The Government lost the baby with the bathwater! It ruined many peoples lives. Fortunately I'd been back in the UK a few years when the blow fell.

After the NZ Government set up the new DoE; the two Engineers in the Research Section working on Energy transferred across to the new Department. My boss, Hank Bauer, the Senior Research Engineer told me privately he would eventually move over, but I could take over his job! I said I didn't know anything about energy systems. Hank said I didn't need to know anything about it.

I thought NZE would want a degree man for that job, and there would be plenty of Kiwi's around who would fit the bill. As Earthquake Engineer, I had developed strong links with the MWD, DSIR and the Universities. Administratively I could have done the job, but I thought it needed a degree person with knowledge in that area.

Brits working in the Boiler Section of the Design and Construction Division, when they emigrated from they UK had joined an expanding New Zealand Electricity Department expecting further conventional steam power stations after Huntly(the show case latest steam power station) leading eventually to a Nuclear Power Station. After it was announced there was too much power generating capacity, and no new power stations would be built for the foreseable future, these lads said to me, there would be redundancies. They got a job with a Boiler Design and Manufacturing company in Perth, Western Australia, sold up and left. I wasn't sure where my job was going, but I thought my knowledge and experience of aseismic design of Power Station and Switchyard Equipments would keep me busy within NZE. Although job expectations would be curtailed, it was, if it was allowed to continue, still a fascinating experience.

We had been rushed into buying a house when we first came over; we did OK, but we hadn't the time to really look around properly; but more on that later. Our son thoroughly enjoyed going to Wellesly College at Eastbourne around the bay from Lower Hutt, and opposite Wellington. Our daughter didn't settle at Chilton St, James School, in Lower Hutt; the sister school to Wellesley. So far as I could see, there didn't seem to be the opportunities for the children I'd expected to see; but then when we got back, there wasn't much going on in the UK either. In those days cheap Air Fares hadn't reached NZ, so a trip home, certainly when paying school fees for both children and having a small second 5 year mortgage, was some way off.

With question marks about how the job would pan out, not having got the house we had hoped to get, a daughter not settling at school, and a my wife who wasn't enjoying the experience, we as a family had to make a choice, stay or return. If we were going back we needed to be back before the boy lost out in the education stakes and could start with other pupils in the new Senior School. We decided to go back; a good decision as it turned out. If NZE had a new power station build program and was preparing for nuclear power, my job would have developed in status; been upgraded; I would have stuck my heels in. Salary increases would have followed automatically.

With the enhanced income we as a family could have bought ourselves into a very nice residential area and home. Something worth staying for.

What was the New Zealand way of life like?

A large area of the Wellington region had been uplifted about 3m by a large earthquake in 1858? What is now Lambton Quay was up thrust above the then water-line. This is now part of the commercial area of Wellington on which Rutherford House and other Government Buildings were built. Geological investigations predicted a similar huge earthquake upthrust could be expected about every 400 years based on evidence of stepped shelved beaches indicating regular upthrusts on North Island side of the Cook Straight. Directly opposite this location on South Island there was evidence that at the same time there were upthrusts on the North Island shore, South Island was experiencing slumping of a similar amount.

A big shake, without upthrust was expected in the Wellington region every 100 years or so. When we arrived it was about 20 years overdue. Nearly 40 years on, the 'big one' has yet to happen, although a year or so ago Christchurch received a big shock that wasn't expected. Christchurh was generally regarded as a less seismic region of NZ.

The Houses of Parliament in Wellington were discovered to have been built across a fault-line. If the fault moved appreciably in an earthquake, the Parliament Building would likely come down. If a precursory shock was felt when the ' House' was sitting; it was vacated very quickly!

NZ Electricities Head Office, built on Lambton Quay was a modern reinforced concrete 14 storey building incorporating the latest Earthquake Building Code Regulations. The Research and Earthquake Engineering Section was located on the $7^{th}$ floor. The first thing I noticed in an earthquake was the curtains moving; then there was a rocking to and fro motion. Fun at first, but you soon wanted it to stop, which in the earthquakes I experienced it did.

Trains were stopped and lines and bridges checked. When given the 'all clear' train services resumed. An earthquake's associated ground movement effected balance for a while, but you got over it after a while; but disconcerting at the time. You get used to it. Usually you feel two shocks: the first or primary wave followed a few seconds later by the secondary wave which causes most of the damage. The time between the fist and second shock is an indicator of how far away the epi-centre is. The time between shocks in seconds is an indication of how far away the epi-centre was. In a big earthquake there are usually damaging aftershocks that can occur days later.

Cars usually lasted longer than in the UK; less rust; they were also expensive and held a good second hand price. There was the general range of family cars you would see in the UK, plus American 'compacts', big six cylinder or V8's. Australian Holdens etc. Good for towing boats and caravans. We bought a second hand low mileage 1600cc Avenger, metallic grey with whitewall tyres; it looked the business and had a purposeful growl!

Bought it from the Main Dealer in Lower Hutt. Standard practice to fit a towbar to discourage shunting when parked up! Kiwi drivers! I remember coming around from an anesthetic after having a wisdom tooth removed and telling the Dentist I wanted a big Holden, when I'd put the fear of C...t up the Kiwi drivers! He took it very well!

Houses were mainly built of wood; wood framed with outer wooden lapped planking, plasterboard inner lining, modern houses having fibreglass insulation between. Rooves were corrugated steel; clay tiles were frownd on because of the weight, which was dangerous in an earthquake. They accentuated earthquake response and could lead to building collapse. Sometimes, as an alternative to corrugated iron, pressed galvanised steel 'roof tiles' with baked bitumen and grit would be used. They were very realistic.

Apart from painting the corrugated roof, roof maintenance also required you clambered over the roof knocking down all the nails once a year so the seal between the rubber washer and nail head was maintained; otherwise you got leaks! The expansion and contraction of the corrugated steel roof combined with wind loading loosened the nails, which is why they had to be hammered down from time to time. Rooves were low pitch so it was easy to walk around on them for painting and other maintenance work.

The older buildings were a direct copy of the Victorian style, but in timber; even Wellington Cathedral and Old Post Office. At first glance the wood had been carved to represents large stone blocks!

Why build in timber? Timber construction would allow shaking of the building in an earthquake with little if anything other than cosmetic treatment afterwards. With unreinforced brick buildings, walls would crack and sometimes collapse. Some modern homes were timber framed with an outer single brick veneer wall.

A proprietary type of home was the 'Lockwood'. This was made from tongue and grooved 4x2 timbers covered on the external 4 inch surface by aluminium sheeting with a baked on white enamel finish. The interiors were varnished. All wiring was hidden by drilling vertical holes down through the wall timbers. You needed to get your electric sockets sorted before the roof went on! The houses used aluminium framed external doors and double glazing. They were very rigid, well insulated buildings with excellent earthquake damage resistance, but it was an expensive form of construction. So, Lockwood houses tended to be smaller to save cost. Modulock was alternative cheaper all wood type of construction assembled from factory built modules. Some houses had a timber load carrying core with a brick veneer wall for appearances; rather like todays timber framed buildings in the UK that are becoming increasingly popular.

Buying and financing a house in NZ back in 1975 was very different to the UK system. There were no Building Society's. Long term loans for house purchase were virtually impossible to get. The NZ Government offered a $25000, 25 year loan to first time buyers and Civil Servants on transfer. The NZ Civil Servants Association offered a $5000 second mortgage over 5 years, and anything extra you loaned from a Bank or your Solicitor over a relatively short period, say 5 years, at which time you had to pay back what was still owed, by taking out a new loan! Well that doesn't sound too bad; but the NZ economy had its ups and downs. Sometimes loans just weren't available so you would have to sell your house, possibly in a downturn in the market, pay back the loan, rent and buy again when money was again available. In that situation you would be unlikely to sell at the best price, and when buying again prices would have risen.

Another whammy was that when looking for a loan, certainly through your Solicitor; his client may want to loan a certain amount for so long at a certain rate. That was the deal; you may need to take on a larger loan than you wanted. There was no flexibility; the loan wasn't tailored to your need!

You can imagine now why we had to buy quickly when after a change of Government, because the NZ Government First Buyers Scheme was heavily oversubscribed, the Government were going to stop further borrowing on the Scheme. We had to buy quickly. Not so bad surely, you could sell and take the Government loan with you to another property? No; not unless you were moved as a Civil Servant under transfer. For any other reason; say health; pollen from the 'Bush' for instance; that had to be considered by a 'Committee'. So, you could be stuck.

The 'Land Agent' played a similar role to that of Estate Agents in the UK, but with some important changes. In the UK, if you are new to an area you call in to see the local Estate Agents, come out with a sheaf of properties to look at and most probably a map of the area. Drive yourself around and get a feel for what's available, maybe arrange some viewings. And repeat until you find what you want more or less.

Not so with the NZ Land Agent. When you visit, he wants to know how many bedrooms, garage, small or large section (plot), area, price range. He will then look through his folder; suggest area for the size of property you are looking for; pick out three options; make appointments to view and take you around in his car. One of the three may be a possible on the short list at that stage. The Agent expects you to buy one of the properties he's shown you; and gets upset if you aren't interested, as he says 'I've shown you what you asked for, why don't you buy!' That takes up a morning of everyones time. In the UK you would have got lots of places to see during the morning's tour of the Estate Agents, which you could have driven by in the afternoon and next day.

The house buying process wasn't easy; difficult to spot a good deal, but worse you were going blind or that's how it felt. Eventually we found an Agent prepared to spend time with us and we got the best we could in the time available to us.

Our intent was to rent for 18 months, look around, find out about the areas and prices, then buy. We didn't have that option, and I think that was a big reason why the family didn't settle.

One good point for the seller was when the purchaser made an offer, they had to state exactly how the purchase was going to be financed, and show proof to the Agent.

The Buyer signed the offer, which the seller countersigned if he was happy. The Buyer then had a given time to get his solicitor to do the legal work and get the finance set up. From memory it was something like 20 or 30 days. After which time the seller was released of any obligation and you as the purchaser would have to apply for extra time to complete the purchase. If the seller thought there was a better offer in the wings and decided to go with that, you bore the legal expenses so far incurred. To establish stability at the time the offer was made and excepted; both the buyer and seller had to deposit a certain amount, something like $2000 which was forefeit by the party breaking the Offer Agreement.

That's how it worked, a bit like wading through mud!

Buying a car was another interesting experience; you would be looking around a car yard; the salesman would come up to you and ask if there was anything you would like to buy? And feign surprise if there was nothing you wanted. He'd say 'What nothing with all these cars! He'd try and take you for a Teest Drive. One of my young daughters favourites questions; lets go for a teest drive. Teest is how the Kiwi pronounces test. It's also peen instead of pen! So when you are first asked if you have a peen, you ask what sort of pin? A safety pin or a round headed pin? Puzzled Kiwi! Guess what a beed means; it's a bed. Lesson in Kiwi over!

Christmas in NZ took place mid summer; hot days, sunshine, blue skies, sometimes windy being Wellington, but outdoor beach weather, not cold and probably wet like the UK.

After nearly 5 years away, it was decided we'd return to the UK. If we were going we had to return fairly soon so that our 10 year old son didn't loose out in the UK Education system. It wasn't a good time to return from the financial point of view; house prices in NZ had remained almost static since we'd bought, and if you really wanted to sell, a loss was not unusual. House prices in the UK since we had left had just about doubled and the NZ dollar fallen in value. But Hey if we were going back it was now or never.

I wrote to Bernard Grant (Rolls-Royce Stress Office) regularly at Christmas and kept him updated with my adventures, and he would write back telling me how things were with the firm. In lieu of new orders and defence cut-backs there had been big redundancies in the company including the stress office. He was busy filing away all reports and technical papers. It wasn't long before he took early retirement. He had moved from a bungalow in Patchway; near the Offices to another bungalow overlooking the Bristol Channel at Portishead; famous for its Radio Station, about 15 miles from Patchway.

We put our house on the market, and when we had a contract to sell and things going well, handed in my notice; 3 months; sold up a lot of the stuff we weren't taking back and rented a house in Eastbourne to complete my notice. The last month I worked for free! Whatever; I wasn't going to break my notice period. As a precaution a job would be held open for me in NZE at Engineer level should we want to return; The Earthquake Engineers job had gone, I would have had to fit in where I could.

We flew out of Wellington ANZ 737 to Auckland and boarded an ANZ DC10 for Los Angeles where we had a two night stopover visiting nearby DisneyWorld; fabulous place, we all loved it.

Then BA 747 to Heathrow, then a hire car to Southampton, over to the Island where we stayed for a few days with my parents and a niebour while we sorted out accommodation longer term until I got a job and we could buy.

# 10. Working In The UK

I managed to obtain full time employment as a Stressman/Structural Designer when I got back to the UK, but the jobs were never that secure, nor had they the zing of the Earthquake Engineers job, which was at Designated Engineer; management level; quite a coupe for me.

Before we left NZ I sent off CV's to a number of firms and contracting agencies, and I made contact again as soon as I was back on the Island. I applied to Plessey Radar at Cowes, on the Isle of Wight, was interviewed and offered a job. I took it, but nothing much seemed to be going on there; I didn't feel very comfortable somehow.

I got an offer a few days afterwards from British Hovercraft Corporation (BHC), at East Cowes, nee Saunders-Roe where I'd started my apprenticeship. They were keen to have me back, and offered me a Senior Stress Engineers position working on Hovercraft structures. I pointed out at the interview that I wasn't an Aircraft Stressman, I'd been on an Aircraft Stress Course about 12 years before, so I would need to be brought up to speed. They said they would train me up, saying I would soon get into it. I was overjoyed to return to what I considered was my 'old firm' and in the stress office too! But it wasn't what it seemed!

Since the Saunders-Roe days, the renamed BHC had been taken over by 'hard men'. A former apprentice Baz. Came up to me and told me that. He was on contract and said he wouldn't want to come back on permanent staff. Very good advice, which I didn't fully appreciate till later events. In my career I worked for two firms wh prided themselves on being run by hard men! My experience of this style of management was that staff went into 'safety/ sloping shoulders mode' to protect themselves. That hard management style by fear was counter productive. Much better to respect and support the highly qualified staff; they responded to that and took collective responsibility with a 'can do' approach to any problems that may crop up; almost inevitable with a fast moving tight budget design and build program. The worst company in that respect I came across was Marconi Space Systems, which stood out in this respect.

By the 1980's Thatcher's Moneytarism Policies had decimated British Industry and with that, design and craft skills were being lost. Industry was seen as risk prone because of the complexity of advanced engineering projects, the performance, and delivery clauses in the contracts which if triggered could be punitive.

Banks were seen as safe, and gave a consistent and better return on capital. To some it must have seemed it would be ever thus.

It was said Lord Weinstock would have made more profit keeping his money in a bank account than investing in Industry. But his wasn't such a short- sited policy; he spread the risk and also invested in high technology for the future.

# 11. SENIOR STRESS ENGINEER @ BRITISH HOVERCRAFT CORPORATION, EAST COWES.

Ray Wheeler was Project Director and Chief Designer; he'd made his name designing the Hovercraft. Albert Week's was his deputy. Ray took on the role of 'good cop', to Alberts 'bad cop', is how it came to be seen by me in future 'discussions' with the pair. Beware the 'good cop'.

I was put to work with John Dadswell on stressing the firm's new AP 188 Hovercraft design. Instead of using traditional aircraft riveted construction techniques and marinised aircraft gas turbines, the new design was to be of all welded aluminium construction, in place of riveting and use low weight high power diesel engines. This would dramatically reduce cost. Aircraft stressing is quite complex, not something you can pick up straight away. The firm seemed to have forgotten they were going to train me up in this. I reckoned it would take up to 18 months of good tutoring and experience to become really proficient in Aircraft Stressing, I don't think any competent aircraft stressman would disagree on that given my general stressing experience. I needed retraining, simple.

We were able to rent a house from BHC at a peppercorn rent which we eventually bought.

My next job was stressing the Search & Rescue Winch installation on the Sea King helicopter; the type of stress work with which I was familiar. And which was appreciated, it was a tricky job.

Other work going on in the Stress Office at the time was stressing the wing centre section and empennage for Shorts latest and last variant of the Skyvan, together with general support for Westland's helicopters. Not really a busy place. The other stressmen I recall were Colin Arnold and Norman Nichols; whose catch phrase when he walked past my desk was 'Know much!' His father had been

Chief Stressman and Norman had gone to a fee- paying Grammar School. He eventually became Chief Stressman some years later.

General resentment was the feeling of the office; no one was really happy there! Here was I an ex-craft apprentice, been around the world, done well, and now given a job in their prized stress office, the pinnacle of their experience; never been anywhere else. Obviously, I could do things they could not, they knew things I didn't. They didn't want me taking any goodies, that may come down from on high because when trained up I could have been a contender! I was glad to be back with my old firm, and would have happily stayed if I'd been left to settle and find my feet.

It would need time to retrain in aircraft stressing. The firm wasn't very busy; the new Hovercraft was the only new project, and the Stress Office was not a very happy place. People I think were concerned about their future; I don't think pay rates were good. Living on an Island they were to some extent considered captive!

The firm was just drifting along. It was headed for a major restructuring in future years. Senior Management must have seen problems coming. I couldn't see a long- term future with the company, not without a lot more work where they would need an extra hand and be pleased to train him up.

I was still getting offers from agencies I had contacted when I first got back to the UK. One firm in Bristol was looking for a Senior Stress Engineer to lead their Stress Team. My Earthquake Engineering experience was of particular interest as the firm was designing a Nuclear Fuelling Machine for the new Torness and Heysham Nuclear Power Stations. I went for interview and was offered the job with a good relocation package.

I decided to take the offer and move up to Bristol. Not as easy as it sounded; housing costs were the main problem. There were also family issues, which in combination with the housing, in the end, didn't make a permanent move viable.

# 12. STRESS TEAM LEADER @ STRAUGHAN & HENSHAW, BRISTOL

Straughan & Henshaw at Bristol was a very well established and respected company whose business was Mechanical Handling. Designing and manufacturing Weapon Handling Systems for the RN Nuclear and Conventional Submarines, Nuclear Fuel Handling and Storage for Nuclear Power Stations and Material Bulk Handling Equipment; Railway Wagon Coal Hoppers; Carousel Wheels etc.

The company was founded by Robert Price Strachan and George Henshaw in 1879 specialising in the manufacture of paper-bag making machinery operating from Lewins Meade, Bristol.

E S & A Robinson bought the business in 1920.

In the 1950's as part of a consortium including: Clarke Chapman, Head Wrighton, C A Parsons & Co and Whesso formed the Nuclear Power Plant Company (NPPC). They were awarded a contract for Reactor Mechanical Plant at Oldbury nuclear power station and subsequently similar contracts for Heysham 2, Hinkley Point B, Hunterston B, and Torness nuclear power stations.

In 1966 E S & A Robinson merged with John Dickinson Ltd to form Dickinson Robinson Group.

In 1972 the company became the prime contractor for weapon launch systems for the Trafalgar class nuclear submarine fleet.

Roland Franklin of Pembroke Associates acquired Dickinson Robinson Group in 1989, and went on to sell Strachan & Henshaw to Weir Group in 1990.

In1998 Weir Group acquired Bridgetest Holdings Ltd, a Manchester based engineering company and its subsidaries: Cunnington & Cooper Ltd, Nuclear & General Engineering Ltd, and Wingrove & Rogers; all well established businesses in the UK nuclear industry.

In 2000 Strachan & Henshaw sold its materials handling equipment division to Swedens Svedala.

Babcock International Group acquired Strachan & Henshaw in 2008. It was subsequently fully integrated into babcock and the Strachan & Henshaw name is no longer used.

When I joined, the company were designing Nuclear Fuel Handling Equipment for the Heysham and Torness Nuclear power Stations. A new design requirement was earthquake withstand. I was the ideal candidate, so I got the job. They were also busy designing underwater weapon launch systems for the Royal Navy's submarines. The firm was also heavily involved in the Bulk Handling business. A very busy company; especially in 1980; when so many of the countries leading engineering concerns were laying people off.

My boss was John Parker, MOD Project Director, an HNC man and ex apprentice of the company. Sharp as needle and decent with it.

I looked after the Stress Team responsible for stressing weapon handling systems for working loads and shock, Nuclear Engineering including Earthquake withstand, and Material Bulk Handling Equipment.

We worked on an invitation basis from the various Project managers. I would be called in by Project, make an estimate of what needed to be done including analyses, hours and cost. This would be agreed. I would monitor progress and if necessary get extra time. It was done in a gentlemanly way. In a QA review across the company the Stress Team came top! I must have been doing something right.

The Nuclear Handling equipment was an interesting experience. When in the Research Section in New Zealand, one of my tasks was to read Earthquake Engineering papers from the Universities etc. I came across a paper from someone working on Nuclear Engineering at Barnwood Gloucester. He had written a paper describing a process for scaling world earthquakes to a credible UK earthquake.

He'd chosen the Indian Pacoma Dam Earthquake record as the basis to demonstrate the method. This quake was most uncharacteristic in that it only had one large pulse; whereas quakes usually have a series of pulses. Because of it's single pulse characteristic, it was relatively easy to scale; so the Pacoma Dam record was scaled to what might be expected in the UK for average conditions; whatever they may be! Remember this was the first time UK Nuclear Power Stations had to be designed against Earthquake!

At Straughan & Henshaw I recognised the resultant response spectra. And had a copy of the paper describing how this design spectra had been devised. Without other modification this had now become the UK Design Earthquake Response Spectra for use in both Heysham and Torness Nuclear Power Stations. Interestingly one power station was built on granite, the other on clay. The ground and local geology has a big influence on the response spectra, granite and hard rock makes earthquake response spiky. It has a greater intensity over a smaller frequency range. The effect of clay reduces the intensity and elongates the response spectra.

A design response spectra needs the input from a number of recorded quakes modified to represent local geological conditions.

I mentioned this to my boss; John Parker and wrote him a note referencing a number of papers on the topic. He took it up with British Nuclear Fuels (BNF). They agreed with what I had said, but said their response spectra was conservative.

As an Engineer I would have liked to know more about how they reached that decision. Virtually all advanced countries liable to earthquake have their own Earthquake Response Spectra; derived for their specific geology, or adopt another country's aseismic design criteria and Codes of Practice. BNF adopted the ASME (American) Codes, but not a modified Response Spectra. I would have thought that to be one of the first comparators. However, this was above my pay grade. People making that decision had honours degrees and PhD's aplenty. I'm sure it was conservative, but not the way I'd have gone about it, that's all.

When I first started at Straughan & Henshaw, I met up again with Terry Hares, previously Weight Engineer at Rolls Royce Bristol. He told me that about 18 months after I'd left to emigrate to NZ, there were big redundancies effecting the whole Division; Manufacturing, Design including of course the Stress Office. At about that time British Aircraft Corporation (BAC) at Filton were also getting rid of people, so nowhere local to go to. The redundant Stress Engineers went Jobbing at Boeing working on the new 757 and 767 twin engined jet airliners.

The Boeing 757 was a new mid size narrow body twin jet airliner, intended to replace the smaller tri jet B.727 on short and medium routes. The B 757 was produced in two versions from 1981 to 2004 when 1050 had been built for 54 customers. Passenger capacity was from 200 to 295, and had a range of 3150 to 4100 nautical miles depending on variant and payload.

After regulators granted approval for extended flights over water, ETOPS, for twin jet airliners in 1986, B757's with suitable systems upgrades began flying intercontinental routes. All versions were powered by Rolls Royce RB 211 or Pratt & Whitney PW 2000 turbo fan engines.

The Boeing 767 was designed concurrently with the B. 757. The B.767 was a mid to large size long range wide body twin jet airliner, carrying 181 to 375 passengers over 3850 to 6385 nautical miles. It was produced in three versions and in May 2015 had received orders for1115 aircraft. The B.767 has ETOPS approval. It is powered by Rolls Royce RB 211, Pratt & Whitney JT9D or General Electric CF6 turbofan engines.

Pilots can be rated for flying both aircraft types which brings economy and flexibility of operation to the airlines.

Design studies to stretch the B. 767 led to the all new Boeing 777.

I later met up with some of these ex Rolls Royce stressmen after they came back from the US and worked for me as contractors at Straughan & Henshaw. Had I not emigrated to NZ, I would have probably been forced into taking the same route; come out of it very well financially no doubt; but would have only come home say 2 or 3 times a year. Not good for family life. Moving the family over was I suppose an option, but that would be disruptive for the children's education, and have eaten into your pay! And there was medical insurance to think about as well.

One evening I drove over to Portishead to see Bernard Grant from the Rolls Royce Stress Office days. Pleased to see him again and catch up. I thanked him for the time and patience he'd given me whist teaching me stressing. I said when I had a problem I often heard him talking it over to me in my mind! True. I think he was tickled by that.

Bernard had taken early retirement, Roy Williams had died, as had the Deputy Chief Stress Engineer Reg. Weatherstone. They weren't old; in their 60's.

A couple of years later when I'd moved to Marconi Space Systems at Portsmouth, I had a reply to my Christmas Card and letter from Bernard's wife to say he had died. I don't think it had been expected. One of those men, who had he had children would have been an outstanding father, of that I have no doubt. A sad loss.

Bernard had been offered a place at University in the late 1940's, but his parents couldn't afford it. As for many others at the time. He took the ONC, HNC and endorsement route to obtain Professional Chartered Engineer status.

You would think I had it made. Well Straughan & Henshaw carried about 50% contractors on their staff! Why? The MOD Design and Development contract for weapon discharge systems for the RN Submarines was finite. Lucrative while it lasted, but once the system had been accepted, the design was unlikely to change significantly for decades. That essentially left Manufacture, which in the future was likely to be put out to tender. Unless there was a new weapon system or special requirement the design team would be run down and people relocated within the company if not made redundant. Career expectations amongst the design staff would be truncated, so they would be likely to move on of their own accord.

Nuclear Power Station work was generally a series of short term contracts; 18 months to 3 years say. The Heysham and Torness Nuclear Power Stations looked like they may be the last to be built for some time.

Bulk Handling: With the demise of UK Mining, their work was mainly from the USA, where they had a subsidiary company. Design was carried out at Bristol. I think manufacture and erection by the American subsidiary.

The firm were very busy and expanding. In the 1980's that was some achievement, most firms were shedding staff.

I could see in a couple of years the firm could be in the same boat. The house I'd bought on the Isle of Wight was an old Victorian Lodge House with a good- sized garden, near the beach, town, and ferry to Southampton. The family had settled down, the children happy at school. I'd bought the house for a reasonable price, £20k as I remember and an affordable mortgage; something not to difficult to maintain if things went 'belly up'. Moving to Bristol would double the mortgage for less of a house!

Doubling the mortgage in 1980's was not a good idea with all the unemployment around. I couldn't see a really long- term continuation of the firm's workload. To relocate to Bristol, they would have had to make an offer I couldn't refuse!

John Parker; my boss; once said to me, when you start running out of work it's too late; he must have had similar thoughts.

Commuting weekends with a young family isn't a good idea. You become detached somehow. I'd been doing it for 18 months, enjoyed the job, but couldn't do it for ever.

I had lunch in the canteen with Douglas Hunt (Dougie); an ex Battle of Britain fighter pilot. He was apprenticed to the Bristol Aeroplane Company before the war and joined the RAF Auxilliaries because he wanted to learn to fly. He attended the annual Battle of Britain commemorative dinner in London every year and met the big names. He was an interesting man to talk to. He retired a Group Captain, in the RAF, had his own personal Chipmunk training aircraft for communications. Said he wouldn't have missed the experience for the world. But wanted an easy life in retirement and settled on a relatively quiet job with Strachan & Henshaw in the office. I guess he'd had enough excitement to last a lifetime. He used to joke, he needed a new set of hydraulics! I guess a lot of us oldies could say the same!

About this time I was approached by an Agency to see if I was interested in a position as Work Package Manager at Marconi Space Systems, Browns Lane, Portsmouth working on satellite payloads? It would mean a pay rise and promotion. I would also be able to commute daily from home. Seemed too good to be true! It was.

A few years after I left, the company had a change in ownership. Around that time John Parker left. To me it signalled organisational changes within the company. It was probably wise not to have moved up to Bristol, and doubled my mortgage.

# 13. PRINCIPAL MECHANICAL ENGINEER @ MARCONI SPACE SYSTEMS, LIMITED, PORTSMOUTH.

I went for interview, saw the Chief Engineer, Ron Turner and the Chief Mechanical Engineer, Ted Clark for whom I would work. Ted was an ex-Saunders-Roe Apprentice, of my intake. I couldn't place him at that time, but knowing the spirit de corps engendered in the apprentices, thought that augured well, he would look after me I felt sure whilst I settled in and found my feet. I got that wrong! Looking back, I think BHC queered my pitch before I got started. The inner circle and all that; there were connections; say no more.

Ron Turner, the Chief Engineer, was an ex-REME Sargant Major. He was working at BAE Stevenage before taking up his post at Marconi.

The Marconi brand was used for the GEC Defence Business, i.e. Marconi Space and Defence Systems formed in 1969, and still known by that name when I joined. It was later renamed Marconi Space Systems Limited.

Sir Peter Anson, a retired RN Admiral was MD whilst I worked for the firm. He injected a dynamic throughout the company. I thought he was inspirational. In WW2 as a midshipman in HMS Prince of Wales, he'd been taken prisoner by the Japanese when his sip had been sunk by Japanese aircraft near Singapore; there was no RAF aircover. The RN task force had gone to intercept reported Japanese landings which proved false. Both Prince of Wales and Repulse were lost.

All staff at Marconi were aware of budgetary constraints and program and tried hard to meet these requirements. It was common for staff to juggle their holiday between program commitments where they could so program didn't suffer! In fact, when setting out your workload, holidays were not included! It was expected you fit them in when you could!

It was a stimulating work environment. However, with a decent fair-minded manager, it worked quite well, but the system could be and was abused by any Manager who wanted to bully and throw his weight around. I like a bit of a challenge in a decent benign atmosphere, but abhor the 'political blame game style' of management that was often the Marconi way. But provided the program was met and on budget, this style of management was allowed. There are other, better ways of motivating a workforce which gets more out of them too.

Matra Marconi Space (MMS) was established in 1990 as joint venture between the space and telecommunications divisions of the Legadere Group (Matra Espace) and the GEC Group (Marconi Space Systems). The merged company was announced in December 1989.

In 1994, MMS aquired British Aerospace and Ferranti Satcoms from administration, bringing satellite ground station components and systems technology to the group.

In 1995 GEC bought 45% shares in the National Remote Sensing Centre for the company.

British Aerospace regained an interest in the company when it merged with GEC's Marconi Electronic Systems to form BAE Systems. In 1999 Filton site was closed and over 400 staff transferred from Bristol to Stevenage.

In 2000 the company merged with the space division of Daimler Chrysler Aerospace AG (DASA) to form Astrium.

When I joined, Marconi Space Systems was overstretched. It had the MOD Skynet 4 Military Satellite payload, L-SAT, INMARSAT and coming up in the rear the new UNISAT project which I was to work on. That was the order given to sharing the stretched technical resources. A new boy was going to find it hard to get a look in!

UNISAT was to be a new communications satellite beaming signals and TV between the USA and Europe. The BBC were an interested, but as yet uncommitted customer. It depended on them getting funding and no doubt tying up users.

Apart from the Project Manager, David Bartlett and key Engineers in the Mechanical Engineering Group, the company had to recruit many Engineers competent in their own discipline, but in the main without space or satellite experience.

One exception was a highly experienced Antenna Manager; Roy Hathaway. He had just returned from Canada and slotted in perfectly. I worked for him on the project as the Principal Mechanical Engineer responsible for the antenna mechanical design; reporting to Ted Clark, the Chief Mechanical Engineer running the Satellite Mechanical Engineering Group (SMEG). Ted reported in turn to the Chief Engineer, Ron Turner. I had in fact two bosses; the Project Antenna Manager Roy, and the Chief Mechanical Engineer Ted Clark. I was part of SMEG, assigned to the UNISAT project.

Good wages were offered; which put some of the firms experienced staff's noses out of joint. The newcomers were often on higher salaries, just to get them in the door. In time they would be levelled down by having lower rises. That's how it generally works!

When I started, the UNISAT project was in advanced stages of negotiation, but it hadn't gone firm. I had no Mechanical Engineers allocated to me to work with, which I thought strange. And Ted wasn't interested in involving me; he was I found out later, holding meetings with the Section Leaders; Weights, Stress, Dynamics and Thermal regarding UNISAT (my project) without me knowing about them and being there! So, if I wasn't being involved, how was I to do my job and co-ordinate? Very difficult unless I was clairvoyant!

Roy Hathaway said to me once 'Ted doesn't seem to be giving you any support Peter, I feel sorry for you'. Roy had been a Technical Apprentice at F G Miles Aircraft at Woodley Aerodrome, near

Reading. He read Electrical and Mechanical Engineering at Reading University and had a degree in both subjects; a pretty sharp turned on bloke. He said when F G Miles Aircraft went bust after the war, the bottom dropped out of his world. He went to work at Shorts in Belfast on the design of the 4 jet Sperrin Bomber, an interim type, before the V bombers would come into service some years later. His next move was to de-Havilland, as it then was at Stevenage. After cancellation of Blue Streak, he moved to the USA and Canada where he again worked in the Space Industry. He'd come back from Canada and joined Marconi Space Systems a few months before I joined the company.

Roy, the Antenna Manager was aware of the log jam, so I thought he would be able to sort it out at Project Manager and Chief Engineer level if he was really concerned.

I didn't see there was much more I could do at that time, other than have a bust up with Ted, or go above his head, which would amount to the same thing. My view was 'early days', things will settle down when the project goes firm; don't make waves.

An interesting insight as to how the company operated is quite illuminating!

Ted couldn't bring in extra staff to work on UNISAT without Dave Bartlett; the Project Manager's agreement. Without that, had Ted brought in staff and put them to work on the project Ted himself (and I) would be in trouble. Surely it was up to Ted to discuss the workload with me, the Antenna and Project Managers and get agreement. If these meetings took place, I wasn't told about them!

The company was severely over stretched technically. UNISAT was declared 4th in the pecking order of priorities; which mean't in affect we didn't get any engineering staff!

A consultant, Ralph Robinson, was parachuted in from Ford Aerospace in the United States to clear log jams and get things moving. After a swift review of other projects, UNISAT caught his attention. He stayed with it for the remainder of his consultancy with the company.

The UNISAT Project was housed in Portakabin's on the Brown's Lane site. The Drawing Office and Mechanical Engineering Group who gave technical support were in permanent office accommodation on the same site.

Ralph made an Executive decision, and moved the UNISAT Project off site into office accommodation in Warrior House, situated on The Hard, outside Portsmouth Dockyard. A shuttle bus between there and Browns Lane operated twice a day, morning and afternoon.

If trying to get technical support was difficult before, it was nigh on impossible now! All Marconi needed to do was get technical staff, the job would have settled down and everything would have progressed in a satisfactory manner. Ted needed to be authorised to staff up the project. A gentle word from the Chief Engineer would surely have resolved any difficulties. I didn't need to know about it; it should have been done through project. It's called Managing!

Getting staff; which they had to do anyway, would have been a lot cheaper for Marconi and no doubt the project, than bringing in a high- level consultant; and with all the disruption it caused taking peoples eyes off the ball, sloping shoulders and watching their backsides instead!

When the project moved to Warrior House, I would go directly into Browns Lane in the morning; get Draughtsmen and Mechanical Engineers started, catch the shuttle back to Warrior House. And get on with my other work, preparing test procedures and so on. A couple of day's later I'd go into Brown's Lane again first thing and check on progress! Nothing had been done! I was told an hour after I had left, a high priority job had come in from another project; ALL projects had greater priority than UNISAT; my work had been set aside and the higher priority work took precedence. But no one phoned to let me know; it was infuriating. My work wasn't making progress.

On return to Warrior House, I reported this to the Project Manager, David Bartlett, who proceeded to kick ass rather than try and resolve the matter by getting extra staff. What was wrong with him? He used to go red as a Turky Cock; someone told him he should see a doctor! Clearly, he was under great pressure.

The Consultant took decisive action. He recruited technical contract staff and based them with the project at Warrior House; a no brainer; all too late for me! That should have been done before isolating the project from direct technical support at Browns Lane.

Needing to be seen to be dynamic, having a grip and making progress I suppose, the consultant moved me sideways. One of Ted's protegés Bill Hardy, was moved in to do my job.

All the technical resourses I'd needed, suddenly appeared on site with him, draughtsmen, stressmen etc. Work proceeded apace. But interestingly the change wasn't made until Bill was in place, everything was set up to go. Magic! Clearly, I'd been side lined and set up as a stuky!

Brilliant man, that consultant! Certainly, knew how to get things done; 'like hell he did'; as John Wayne might have said! He screwed things up for me big time. I should of course have got close to him, but as a newcomer in the firm doing that would have made enemies. I had to rely on his sense of fair play, decency and deal with the cause of the log jam with integrity. Too much to ask? It certainly wasn't me; but I was eminently blameworthy! He never spoke to me about engineering support, if not me as Principal Mechanical Engineer; why not? That in itself sent a signal and undermined me.

Had the consultant from Ford Aerospace not been brought in I think everything would have gone well. I was particularly good at co-ordinating, and although lacking in satellite mechanical design had a remarkably wide experience needed for the job.

Basically, Management had cocked up; I was clearly expendable, a newcomer to the firm with no street credibility. I was the ideal patsy to protect everyone else in the 'Blame Game Management' style of the company. It happens to the best of us; get over it!

Roy Hathaway to his credit insisted I reported directly to him on the project; not demote me to work under Bill Hardy. I think that made the sideways move tolerable. Roy later told me the Consultant thought he had been a little hasty in moving me! Too late then; I'd been stuffed. My philosophy was keep with it; it will pass; when Roy moves to another job perhaps he'll keep me with him. If it was up to him, I'm sure he would have.

Before the consultant left to go back to the US, he called me into his office, and said to me 'If some ones out to git you, they'll git you, git out'! I replied I'd left a good job running a stress team at Bristol to do this job; I've recently brought my family back from New Zealand; there is nowhere for me to go! I have to make the best of it. He then said we don't do things like this in the states! I often wonder why Ralph didn't see me when he first arrived; had he been briefed against me, was he protecting someone and just carrying out orders? I'll never know; but that was the Marconi way!

He had now perhaps seen things he didn't like and had time to reflect. Too late; he'd put a knife in my back, and by inference associated me with his criticism of Ted Clark! I had been totally loyal to the man, Clark had shafted me!

I found out later the consultant in his report had damned Ted Clark, my boss, for not providing me with technical support and recommended he was removed from the post of Chief Mechanical Engineer. I had been careful not to criticise Ted, but to anyone taking an interest, Project Manager down, it was clear what was wrong. I and others could see the train crash coming. It would have been suicide for me to criticise my boss, I never did, but he must have believed I had from his subsequent actions. Although I think Clark was trying to protect himself; that would have been clear to a 10 year old! If the consultant had put his weight behind me; things would have been fine.

Thinking about it, it wasn't in some people's best interest for me to be a success; I was their alibi! If you've been in this situation, you'll understand; otherwise you'll think I'm having a whinge! But read on my friend!!

Later, UNISAT was cancelled, the BBC couldn't come up with the money we were told. The team broke up and I was tasked to take over the SMEG Mass Properties Group, because the current Section Head was leaving.

I had to complete writing Procedures for measuring Centre's of Mass and Moments of Inertia using Schenk Mass Property Measuring Machines and generally finish setting the section up.

It was also my job to weigh and use these machines to measure the centre's of mass and moments of inertia of 'black boxes' and Antenna.

I recommended that the full process of measuring mass properties was only undertaken on the 'first off', subsequent units need only be weighed, saving a lot of time and cost. This was approved by all Projects.

I was responsible for the mass property estimates and measurements on Skynet and all other satellite payload equipments designed and produced by the company.

I was invited to attend a TV link at Brown's Lane showing the launch of a Skynet Satellite. It was supposed to be a celebration, a reward for my endeavours. All the bigwigs would be there. I sent my apologies; living on the 'Island' and at East Cowes, the other side of the Island from Ryde and Portsmouth, would have made the event difficult to attend. Go home, change, get back in time, and then getting the ferry back. The last boat wasn't very late in those days. Still it was recognition for what I had done. Looking back, I should have made the effort even if I'd had to stay overnight B&B somewhere and take a taxi! That's what the savvy operator would have done. And that could have made the difference to my career; quite likely someone got upset I didn't go, but then, nobody offered me a bed! That reaction would be typical! The launch vehicle rocket exploded on lift off, destroying the Skynet 4 satellite.

The Mass Properties job was straight forward enough, challenging only in that Projects would want work done as of yesterday; so, you had to keep your eye open, anticipate and pick up the work when you could to fit everything in. It was all unnecessary hassle, but that was the management style, be mean, keep them keen as they say.

I developed Mass Prediction Methods for electronic units and antenna. This required weighing the unit mass of complete PCB's and converting this to an areal density. Chassis were treated similarly, but in this case the overall volume of the basic chassis was calculated and with the weighed mass, a volumetric density was calculated, and so on. A number of PCB' and chassis were used to build up a data base and plotted: area v mass for PCB's and volume v mass for chassis. Best fit curves were plotted and max and min curves drawn for comparison to indicate spread of accuracy. Antenna area v mass was plotted, Antenna attachment was predicted by plotting antenna area v attachment mass.

The traditional approach was to estimate the mass of each part and so build up a mass of the complete unit. The problem with this method was there was no real way of estimating how the final weight would come out. All you had was a not to be exceeded final mass that had been conjured up by Project. Now I had developed a means for predicting final mass, design progress could be monitored with greater certainty of the final outcome. It was a step forward.

This prediction method was tested against the traditional method and found to be more accurate. Clearly the full estimate was needed also to confirm the prediction.

I was given the task of Predicting the Mechanical Engineering Group Forward Work Load; group by group and Engineer by Engineer over all the projects.

I devised a routine and had a programmer put it onto the computer. I updated this monthly.

Life was torrid at Marconi. A stress job came up at GKN Aerospace (nee BHC), in their Commercial Division on the Island. I was interviewed by the Personnel Director, who said if your interested, come back and see me. I then went for interview with two Engineers. One said to me does your boss know you're here for interview? I said no, I've taken a days leave, but if you make me an offer and I accept, your free to contact who you want. He then said it's like that is it! The other interviewer said, that wasn't on, or some such. Clearly Ted had said something! I left it at that! The Island is a small place. Back to work at Marconi.

I was still keeping in touch with Roy Hathaway; who was in a different part of the building. He tried to get me transferred to work with him again; but Ron Turner, the Chief Engineer said no. He had other ideas. I guess Clark was under instruction to let me go!!

When Ted Clark was out for protracted sick leave. I was given £1000 a year pay rise. I thought maybe now things are going to get better and my efforts were recognised. On reflection, not sure if this wasn't part of the overall strategic plan to get rid with no comeback!

There was a change in Personal Manager, a new man was appointed, but within a month or so he resigned! Whilst he was serving out his period of notice I was made redundant; or rather the need of the Mass Properties Section was!

Interestingly, the senior people whom I had worked for had now retired. It was as Ted Clark told me a 'window of opportunity'. I had 3 months notice to serve.

Every month Ted would call me in to see him and tell me he was trying to find a new post in the company for me!! What do you think of that!! I can't use the appropriate word in polite company. He enjoyed every minute of it!! When I was leaving he said if any employer contacted him about me, he would say what he thought!! I thought I was being blocked until I contacted the Marconi personnel manager; as mentioned later.

I came under Mike Algar, who headed up the Dynamics Group, Ted had never carried out his promise to make me Mass Properties Section Leader when I took over the job. Mike told me when he was told the Mass Properties Section was being closed down, I would be given another job. He said he wasn't very happy with the outcome! But he didn't do anything about it! As I said to Mike; neither was I!

The firm then went through the re-location/placement process.

I went up to GEC Whetstone near Leicester for interview. They designed and manufactured large industrial gas turbines for electricity generation. Whilst there they showed me what was left of Frank Whittle's Jet Engine Test Bed. Interesting stuff.

GEC had the feel of Marconi Space Systems. If Bristol was out, Leicester certainly was. I was offered a place at Brown's Lane in the Business Studies Group, as a result of my forward load predictions. I was interested, but was told it would mean a serious drop in salary and demotion! Not acceptable. I took the redundancy, and the best thing that could have happened to me.

On my last day at Marconi, when I went to sign out; I found the firm was trying to short change me of my entitled redundancy pay. I said if you don't pay what I'm owed, I'll go public; we all know this is a put- up job! A phone call was made; I was paid the full amount owing. Some firm! I'd worked for iconic first- rate companies in my career. I can honestly say without a shadow of doubt Marconi was the worst firm I'd ever worked for. I sometimes still get flashbacks!! I got out of the Design Office of BHC at East Cowes before that got really nasty! There was a connection between BHC and Marconi at Management level; an inner circle shall we say.

During my time with Marconi I had been driven close to a breakdown! I took solace by going back to church. Sunday morning at the local church, St, James, East Cowes, and evening service at St. Mary's across the river at West Cowes. Sally and I would go with my mother to evening service; she lived near St. Mary's. She was concerned for me, but pleased to give her support.

I enrolled for Confirmation Classes repeating what I had done years ago. Found great spiritual support from this. I certainly respect anyone with genuine religious belief. What you believe in depends in the main to where you live and the family into which you were born. And who's to say we are not all worshiping a similar ideal, but through a different mantra!

It struck me more than once that some of Marconi senior management had psychological problems!

Going back to before I was made redundant; I tried from time to time to find other work in the general area within commuting distance. Sometimes I got an interview, but then all went quiet!! Until working at Marconi, I would get an interview without much trouble. And then most times an offer! But not know!!!

However, I did get an interview at Vosper Thornycroft Shipbuilders at Woolston, Southampton. It was a long shot. They were looking for a Deputy Chief Structural Engineer. I was told they had a candidate, but it wasn't clear whether or not he would accept. I was given a tour of the Shipyard; busy building GRP Single Role Minehunters for the Royal and Saudi Navies

Just after I'd received Redundancy notice, I was asked to a second interview. The first thing I said was that since the first interview I had now been made redundant! I thought that would kill everything stone dead. I didn't perceive any reaction to that. Redundancy was an annual ritual at Vosper Thornycroft I found out later, but one I survived to retire on my 65$^{th}$ birthday, 13 years later. And in my later years until I retired an annual merit pay rise; usually unheard of in private industry! So, I must have been doing a good job for them!

I was told they had filled the Deputy Chief's role, but they would keep me on file for future reference. A big new order was in the pipe-line I was told.

I wrote asking for expenses for the second interview because it was really for their benefit, no offer came from it and being redundant funds were tight. They sent my expenses.

Whilst serving my redundancy an ex RN CPO who had also also worked on UNISAT with me, came up and said they shouldn't be making you redundant; it shouldn't happen. Someone else said you are the only one who gets back to us if we ask a question and tries to help!

I lost count of how many letters I'd sent out for jobs, 300 or so! Nothing. I concluded Clark was somehow blocking me; as he'd indicated to me, he would when I left! I wrote to the Personnel Manager; pointed out there was a personality clash with my boss, expressed my suspicions and if I found any evidence of that the proverbial would hit the fan. I also asked for a reference.

I got a reference of sorts. It came with a nice letter from the new Personnel Manager saying he wished he'd had time to relocate me, but was busy settling in himself! Maybe further bad mouthing was forestalled by my letter to the Personnel Manager, who knows!

After I had left Marconi, Chris Trumpess, a mechanical engineer in the group went in to see Clark and told him what he thought of him. Chris later left to work on the Channel tunnel infrastructure. I think he was sickened by what he'd seen.

I was relieved to be leaving; I often felt sick and retched when going into the site! Good job I got out; long term the strain would have effected my health. I later developed Type 2 Diabetes; possibly a side effect!

With no work turning up I got offered a place at Southampton Technical College (now Solent University) on a year long Computing Course at HNC level for which I would get a grant and funding.

Having got a banker, I settled down to working on the house laying a path and patio, casting some of the paving stones myself. The next project was to rebuild the kitchen. We have an old house with odd size rooms, particularly the kitchen area. A kitchen kit wouldn't really fit; so, it was buy sheets of wood, door panels etc. and build the carcase to fit.

I had been out of work for some months, Vosper Thornycroft hadn't got back to me, I was not getting anything elsewhere, so I was getting concerned. I saw an advert for Aircraft Stress Engineers at GKN Aerospace. I was interviewed by Norman Nichols who was now Chief Stressman. He said I know you've done some good jobs. I don't understand why Ray Wheeler didn't stop the nastiness at BHC. Norman went on to say that he would offer me a job, but Ray had written to Personnel after I left the first time, that I was not to be employed by the firm again! He went on to say that the firm had been divided into Commercial and Aerospace, Ray was now Commercial Director, so although not now directly involved, there could be a problem. Obviously, there was some come back, because I didn't get an offer. If I see Norman out and about these days, he always looks a bit guilty! I sympathise, I found Marconi put a strain on your integrity in order to survive.

Of Ray; until he realised I was on to him; he always gave a cheery wave and word when he saw me in the town! He also lived in East Cowes. I always laugh at his reaction when he saw me after I had retired driving a 3 year old BMW Series 1; his head nearly spun off his neck! In surprise of course.

The Computing Course took the pressure off, but I was still looking hard for a job; which would have been the best option if something decent had come up.

Golden Rule: If a job isn't working out for you, you can't get things changed; get out. Get out before someone gets a hate on, gets their claws into you and builds up a case against you. I didn't get out; I should have immediately I was moved sideways. It would have meant going to British Aerospace at Bristol most likely and the family would have had to come with me. I'd done my best; it was now their turn. But with children etc. it's never as easy as that. You mostly soldier on and hope things will improve. It's my experience if it's not right in the beginning it won't get better later. Get out!

Maybe I seem to have spent too long with the minutia of life working for Marconi. I spent 7 long unhappy years with the company; driven close to breakdown. This isn't only a story of making my way in Engineering, but a social account of what it was like working on the inside of one of the country's most prestigious company's.

Marconi Space Systems is no more, it was taken over when GEC collapsed. But in whatever new guise the site operates, once you promote a certain type of manager; they seem to proliferate the system. I imagine it is still a stressful, politically inspired place to work under whatever banner!

Its interesting to compare my experience working with other firms where I stepped into a new area and was very successful with that of the BHC and Marconi experience.

The stress office at Bristol Engine Division, was a significant step into challenging technical analyses, where accuracy, attention to detail and application were essential ingredients. I was carefully trained and mentored, so that from a newcomer I ended attracting some of the most interesting work.

At New Zealand Electricity I had to learn a new discipline; that of aseismic engineering. I developed such a grasp of the subject; I organised in house seminars on Earthquake Engineering and co wrote the New Zealand Electricity Earthquake Engineering Manual. And was rewarded by promotion to the newly created position of Earthquake Engineer; which I won in open competition within the NZ Government Technical Civil Service; something of which I am immensely proud.

I'll skip briefly over the British Hovercraft experience except to say the extent of aircraft structures re-training seemed to be confined to working on their new Hovercraft for about 3 months. I heard later the company needed a contract stressman for 3 months work on that project! It was my impression there wasn't much work in the office after that. Perhaps I should add that Aircraft Structure Stressing is a specialised skill that takes some years to develop; and then under the tuition and mentoring of an experienced Senior Stress Engineer. My Powerplant Installation and Seismic Analysis went some way towards this skill. I made clear at interview that I wasn't an Aircraft Stressman. The, interviewers: Jack Ashton (Office Manager) and Albert Weeks (Deputy Chief Designer, or some such senior post) said I would soon pick it up. For some reason the Chief Stressman, Reg Goyne wasn't present.

An Agency contacted me a few months after I started. The job wasn't going well. I was offered a job interview at Strauchan & Henshaw, Ashton Gate, Bristol, running their Stress Team. They were interested in my Earthquake Engineering experience with regard to their Nuclear Engineering Contracts and Shock Analysis on their Weapon Handling Systems for RN Submarines. I accepted the job and resigned from BHC after about 5 months with the firm. Hardly time to have become proficient in aircraft structures retraining had it been ongoing.

For Strauchan & Henshaw I ran the Stress Team, which came top in a company wide QA audit. My Earthquake Engineering experience in NZ was put to good use. I was able, using copies of papers and documents I'd brought back with me, been able to devise a quasi-statically applied loading that BNFL agreed could be used for aseismic analysis rather than a finite element (FE) analysis. The Project Manager wrote me a memo, thanking me and saying I had saved the project about £20k in FE analysis.

After I was made redundant at Marconi Space Systems, I found a job at shipbuilder Vosper Thornycroft. Again it was a new experience for me, learning the design techniques of GRP construction. Responsible for Minehunter Shock Trials instrumentation installation; pre shock hull survey, co-ordinating the shock trials, post shock hull survey; record damage, devise repair and strengthening to be incorporated on all ships of the class.

Design of high speed attack craft shaft line propeller brackets, noted for their lightness.

Responsible for the seismic analysis of Nuclear Fuel Storage Cells.

Transferred to the Research & Development Group I developed a patent for ship outfit which represented a step change, saving considerable cost over the conventional method. I had another idea patented for a steel to GRP joint, and came up with a modular cabin construction concept, that offered a significant step forward in ease of build and installation.

In comparison my treatment and experience at Marconi Space Systems stands out like a sore thumb! Why?

At one stage whilst I seemed to be playing a game of musical chairs, I saw Sir Peter Anson! Not a wise move perhaps. He will ask questions and if enough senior people want to cover their backsides, he won't get very encouraging feedback. And probably find me a backwater where I wouldn't annoy anyone. I think that's partly why I was given the Mass Properties job; not high on anyone's list!!

It's quite clear by this comparison that given the resourses and reasonable guidance I would have done the Work Package Managers job successfully. However, I would have found working on even more demanding projects a strain, which I believe in the long term would have seriously undermined my health. In that respect, I was lucky to get out intact!

The Marconi name is no longer with us!

# 14. SENIOR STRUCTURAL DESIGN ENGINEER @ VOSPER THORNYCROFT (UK) LIMITED, WOOLSTON.

Working under the kitchen floor one day; whist I was otherwise unemployed; replacing floor boards and joists; I got a phone call from Vosper Thornycroft asking if I could start the following day! They said their situation had changed; they now had a new order. I said had you rung a couple of days ago I could have come straight away, but now I have no kitchen or sink and have to get running water on and the oven working at least! I was put onto Alan Dodkins; the Chief Structural Design Engineer. The start date was agreed; giving me about 3 days to get a basic kitchen up and running. Salary was less than I had been getting at Marconi; it was manageable, and the job title had a certain ring to it; Assistant Chief Structural Designer; just short of a designated car parking space!! An annual salary rise was in the pipeline, and it was promised I would receive this increase. About the best I could do in the circumstances.

I found Vosper Thornycroft (VT) a very gentlemanly run concern. People in the design offices and in the yard building the ships were highly competent, helpful and easy to get on with. Great sense of kindly humour too. A super place to work. What New Zealanders would call a bit of 'Old English'. I was to have 13 very happy and productive years with this company.

Vosper Thornycroft was formed in 1966 by the merger of Vosper Limited of Porchester and J I Thornycroft of Woolston Southampton.

John Isaac Thornycroft founded a ship repair firm in Chiswick in 1866; relocating to set up a shipyard in Woolston, Southampton in 1904. The first ship built at Woolston was the destroyer HMS Tartar; launched in 1907. The yard also built commercial ships.

Vosper was established in 1871 by Herbert Edward Vosper at Portsmouth. In 1936 the company moved to Porchester, becoming known as Vosper Limited. They built Sir Malcolm Campbells water speed record breaking Bluebird K4, reaching a speed of 141.74 mph in 1939. It used the same engine as his previous boat K3, the 1937 record holder built by Saunders Roe at East Cowes.

Vosper became famous as the builder of small 60 to 70 ft. MTB's and MGB's for the Royal Navy in World War 2.

Vosper Thornycroft flourished even during the lean times for warship building, mainly through successful sales efforts in exports and diversification into training, support services and nuclear engineering.

Whist completing orders for RN frigates Vosper Thorny croft were designing and developing production processes for glass reinforced resin (GRP) minehunters predominantly for the RN. The first design was the Hunt class minehunter, the largest warship ever built out of this material. These ships entered service in the early 1980's. They had a secondary role as offshore patrol vessel.

The follow on single role Sandown Class minehunter, HMS Sandown (first of class) entered service in 1989. This class was also of glass reinforced resin construction, but took a technological step forward from the partly bolted Hunt Class by being of all bonded construction.

In 1998 VT acquired the specialist military boat-builder, Halmatic at Porchester.

In 2001, in their most ambitious diversification project, VT started work on the US super yacht Mirabella V for former Avis Car Hire boss Joe Vittoria. The yacht was completed in 2004, and at that time was the worlds largest single masted sailing vessel, with a mast height of about 87m and overall length of 75m

From 2002 the overall business interests were known as VT Group plc.

In 2003 the shipyard at Woolston relocated to new state of the art ship building facilities in Portsmouth Naval Dockyard under the name VT Shipbuilders. This move was encouraged to get a share of the Royal Navy T45 Destroyer and follow on Aircraft Carrier work.

In 2008 VT Shipbuilding was merged with BAE Systems Glasgow based Surface Fleet Solutions subsidiary to form BVT Surface Fleet. The VT Halmatic boatyard site in Porchester sold off to Trafalgar Wharf. Halmatic moved into Portsmouth Dockyard Naval Base.

Also in 2008 VT Group aquired BNG Project Services from the government owned British Nuclear Fuels. The Project Services acquisition gave VT Group access to the nuclear industry in the UK and abroad. The business was renamed VT Nuclear Services.

In 2009 VT Group sold its share of BVT Surface Fleet to BAE Systems, becoming BAE Systems Surface Ships.

In 2010 Babcock International Group purchased VT Group. Babcock's integrated the UK part of the VT Group into its own business.

In 2012, Babcock's sold VT Group to the US based private investment fund Resolute Fund 11 LP, linked to the Jordan Group.

VT Group only survives now as a privately held United States defence and services company, with its origins in the former British shipbuilder Vosper Thornycroft (VT). The British part of VT was merged, as mentioned previously, into Babcock International Group and no longer uses the name.

What a game of musical chairs and how sad to lose the name of that once proud company Vosper Thornycroft UK Limited.

However, when I started, this was all a long way off. VT had just received a follow- on order for GRP Minehunters for the Royal Navy; eight new ships in all. The firm was also persuing overseas orders for the minehunter; and new corvette and high- speed attack craft orders.

They were at the time, developing a steel Corvette design to meet the requirements of a Middle East customer in competition with other European Shipyards. An order they eventually won for two corvettes, very powerfully armed warships.

The company were also designing a GRP Minehunter for the Spanish Navy in collaboration with a Spanish shipyard. There technical people were located at Woolston alongside VT designers; who together produced the new design. This minehunter was based very closely on the VT Minehunter design; a technology transfer agreement; the ships were to be built in Spain. Some design improvements were incorporated into the Spanish mine hunter.

This was the first project I was concerned with designing and stressing the main machinery and diesel generator rafts, protecting the machinery from shock and also providing noise insulation required for the mine hunting role. Other work included hull structural design.

My next job was working on a large floating de-gauzing rig to be used for de-magnetising submarines. The submarines floated inside this ringed structure made from GRP which contained the de-gauzing cables.

This structure had to float in the sea for 100 years. The original design comprised a number of GRP pontoons bolted together with hi Nickel corrosion resistant bolts and Phospher Bronze plates. The GRP connecting plate had to be reinforced to prevent the interconnecting bolting shearing though the GRP in a 100 year storm. The design was later simplified, but the MOD decided not to continue with the project and did the de-gauzing some other way. At this point, I feared redundancy, especially being a newcomer with less than a years service. I thought last in-first out!!

I got an interview at British Aerospace, Hamble who were looking for stress engineers. I was interviewed by Colin Arnold; of the BHC Stress Office, who was now Chief Stressman at Hamble. He didn't like the way I had been treated at BHC. He offered me a job. Initially it would be check stressing design changes to the BAE 145 short haul jet-liner. He said that's the finest way to learn stressing, he would put me on various courses to get me up to speed. It was very tempting; it would mean taking a launch trip across the Solent from West Cowes to Hamble; a twice a day service, alternatively by Red Funnel, then by train to Hamble Station, which was directly outside the design offices. Whilst I was thinking about this, he went to see his boss. It was agreed they would make an offer. In an aside Colin said his boss had come from Marconi at Portsmouth; Ted Clark had worked for him, and when he left Ted was promoted to take over his job! I thought no-way; I'm not going to walk into Ted's clutches again! So, I didn't take the offer up. I've always been very interested in Aircraft Stressing, and have the books: Bruhn, Megson, Nui, and Perry. Although I have a good handle on the theory, I lack the practice. Civil Engineering Structures is a different kettle of fish. Most Civil Structures are covered by the text books, not the same with Aerostructures which are more complex. This is what makes Aerostructures so interesting; it's unlikely anyone really knows it all!

I stayed at Vosper Thornycroft and went onto design work for the new batch of Minehunters. GRP suppliers had rationalised their range of GRP cloth on offer. These cloths were used for boat building, baths, cars, etc. In fact, any product made from GRP.

The particular cloth selected and tested used for the design and build of the earlier Minehunters was no longer available. The Minehunter order represented a relatively small amount of glass cloth production.

My job was to re-stress the hull using a new cloth. A wonderful way to find out about the ship and GRP design.

Having stressed a larger HIAB crane seating, it was an interesting experience to oversee the Proof Load Test! The seating was designed for a Proof Load of twice Working Load, with an Ultimate load of 4 times Working Load without failure. Even so the Proof Load is a serious test of design integrity. It passed with flying colours; live to fight another day!!

The interesting thing about ship structural design is that a couple of Stress Engineers, or Structural Designers as they are termed in shipbuilding, can between them design a whole ships structure! There isn't generally the great attention to detail that is needed in aerospace structures to pair weight right to the bone. Designing with relatively thick GRP plating and steel panels designed for sea loads often looks after wrinkling. But 'oil canning' on light steel decks does have to be closely watched. Everything is built to a more substantial scale, including light alloy structures.

HMS Inverness, the second ship of the first batch of Minehunters was scheduled for Shock Trials off Rosyth in Scotland. My job was to co-ordinate with the MOD and VT Instrumentation Group on the location and installation of shock sensors on the ship. It was also necessary to get this equipment on loan in some cases from the MOD for the trials. A helicopter had to be arranged for the company Photographer to film the trials.

Together with a QA Inspector, the GRP Foreman and MOD Overseer, we inspected the ships hull pre and post shock trial.

Pre- Shock Trial, the Minehunter was taken out of the water on a Ship Lift at Rosyth. The hull exterior was inspected for any damage, which if found would either have been repaired or just marked with marker pen and photographed if of a superficial nature and could be left. None was found.

Next the hull interior was surveyed. Ship's staff (the Navy), cleaned out the bilges, then we clambered around with torches, marker pens and camera. Any defect found was marked and photographed and location noted. The sort of things we were looking out for was evidence of de-lamination, shown by a local lighter colouring of the GRP that could occur due to heavy weather. This in itself did not constitute loss of structural integrity; more an indication of local stress relieving, unless of major proportions that would indicate possible de-lamination of a major structural member, when it would need to be repaired.

The hull also passed this survey; any internal defects found were small, they were marked, photographed, position noted and left.

Post-Shock Trial the process was completed, but this time the inspection was carried out with the ship in a large shed to facilitate any repair work needed.

Externally the bow thruster fairings needed replacing, but the hull maintained water-tightness and the ship could have continued to operate without restriction.

Internally further structural de-lamination was found; sometimes new, sometimes an extension of that previously found. These were marked by marker pen, photographed and location noted. Other cosmetic and superficial damage was noted; easily rectified simply by allowing greater clearance between panels etc. to allow for free movement under shock.

Rather than carry out repairs at Rosyth, it was decided to patch the ship up and sail it down to the company's Porchester Yard, take it out of the water on the Ship Lift, replace the bow fairings, repair de-laminations and incorporate simple design modifications to prevent future cosmetic and superficial damage on operations. These design enhancements were incorporated in other ships of the class in service and new build.

This work and extent of repairs was agreed with MOD Overseers. The ship passed the Shock Trials with no effect on operational efficiency, effectively the damage was of a cosmetic nature.

The Shock Trials consisted of the instrumented and de-munitioned ship being mored off Rosyth by the Fourth Bridge. Powerful explosives were positioned in particular orientations and locations from the ship. For instance, on either side of the bow, amidships etc. However, these were located separately for each shock to prevent accidental discharge of all explosives at the same time!

Watertight doors were closed, the crew braced themselves and an explosive set off; the least severe first. On board were Ships Crew, Shipbuilders representatives, MOD representatives.

The company photographer filmed the trial from the helicopter getting a good sequence of the impressive explosion plume.

Immediately after the first test ships staff carried out a thorough inspection of the ship to check for and assess any damage. The ship then carried out manouvering trials and returned to Rosyth.

This was repeated over the next few days for different explosive locations.

Then the ship was taken out of the water on the Ship Lift and into a shed for the Post Trial survey as described above.

Other interesting work came my way helping with the design of a new steel corvette; hull and bulkheads.

My next task was strengthening the Propellor bracket for High Speed Attack Craft of the Egyptian Navy. Extra bracing was required. To establish the loading I developed a Finite Element model of the rear hull.

From that I went onto designing the Propellor Brackets (P-Brackets) for Quatar High Speed Attack Craft contract. The design of P-Brackets and intermediate prop shaft brackets is quite a tricky job. You have to make sure the brackets and hull local stiffness and thus frequency is well out of the way of any resonance at cruising and max. speed with the prop blade rotation for smooth vibration free running. It is almost impossible for there not to be vibration somewhere over the speed range, but this should be restricted to slow speed manouvreing on a High Speed Craft. Minehunters slow speed manoeuvre should be vibration free, or as close as can be achieved so as not to set off an acoustic mine!

Because of my Earthquake Design experience, I was asked to head up the design team working on BNFL Nuclear Storeage Cells that had been won by the VT's Engineering Group; an offshoot from Shipbuilding to try and move away from such large reliance on Shipbuilding. Martin Jay also successfully moved the firm into Fleet Training and other areas.

Getting back to the BNFL Nuclear Storeage Fuel Cells. I was given a contract Stress Engineer to assist me. VT Shipbuilding Drawing Office were also to be involved, so a strong team.

We were given a schematic of what this Nuclear Fuel Storage Facility would be, based on an existing in service design, but the new design and had to meet ASME Code aseismic design requirements.

On the face of it this seemed a reasonably straight forward task; certainly the Engineering Group Management thought so.

There were a couple of problems that had to be overcome first; well three really! We had to get our heads around the ASME Code. We had to progress the initial design on conservative stress levels as advised by BNFL, using the ASME Code in the final design check calculations; a belt and braces approach that provided a safety net to catch any rogue design calculations. A truly independent check, rather than an arithmetical check alone; where everyone goes down the same rabbit hole, in danger of making the same fundamental design error. That's why there is always an independent design check carried out from the final design drawings. Any significant calculational discrepancies are investigated and resolved.

But before stress calculations could be got underway, we had to ensure that should a fuel rod container be dropped when being loaded or unloaded, there was sufficient drop height for the honeycomb dampers to arrest the container before it hit the bottom and was burst open releasing radio-active material. The drop allowed for was insufficient to do this, so I had to design a double acting damper to absorb the energy.

The project resisted this, trying to argue the double acting damper wasn't necessary!

The next problem was that the operating temperature of the fuel cell, caused the steel tie rods to be permanently stretched by the expansion of the casing. I got over this by layering cork between the plates. Correctly torqued up the tie rods contained the fuel cells, but the cork took up any thermal differential by compression; thus, the tie rods remained elastic, didn't stretch, and so remained tight.

The Engineering Group managing the project didn't understand why the design changes were necessary, even though the reasons were carefully explained to them. They were a gung-ho lot, and quite frankly out of their depth, and in panic mode! That, was my and the Drawing Offices feeling. But later the Engineering Group project manager told me the new Fuel Storage Cell design was operating at twice the temperature of the earlier lower operating temperature design the schematic had been based on. And yes, it was necessary to incorporate the design changes I'd insisted upon. A difficult job completed on time under considerable pressure, and also having to work somewhat in the dark.

My contract stress man kept saying 'just wait till the proverbial hits the fan!' That gives some indication of the pressure and difficulties of the job, which by basic engineering were overcome without drama.

Glad to get the job over with.

Soon afterwards the Engineering Group got another BNFL contract for which nthey wanted me tom head up the design. I refused saying they were a bunch of rough necks. I thought I might not get away with it, next time. The Drawing Office were asked to carry out the drawing work. They replied that if design had refused, they weren't doing it either!!

Shortly after this the R&D Manager Dr. Malcolm Courts (Mac) started a new project investigating the design of a modular corvette. I was invited to join the team as Structural Designer, and in particular looking into ship's outfit.

Blom & Voss, the German Naval Shipbuilder had designed a modular warship; using a weapon module for a standard gun, engine room module, auxiliary machinery space module, bridge module and so on, making the ship easy to specify and develop for the customer at reduced cost provided they incorporated these modules.

VT saw Blom & Voss as a major competitor for corvettes size and smaller warships, so decided tom investigate for themselves. It was not going to be easy because each customer seemed to have a specific unique band specific requirement; beam, overall length, armament etc. so it was likely the best that could be achieved would be a set of guiding principles that would adapt to specifics.

The firm had joined The Welding Institute (TWI) study group looking into the use of adhesives and applications. I was seconded to attend monthly meetings at TWI's headquarters near Cambridge. I'd drive up the afternoon before, have Bed & Breakfast, attend the meeting next day and drive back.

The second year the MOD joined and arrived in triplicate! One of their consultants was expert on bonding steel plates underwater to steel hulls. His idea of high temperature was water temperatures of 40 degrees. For the other companies funding this project; undersea bonding temperatures were of academic interest; we needed every day adhesives that could withstand temperatures of 80 degree C plus exposed to the elements! Unfortunately what had been progressing quite well; there were representatives from Shell and other large commercial undertakings; took on a change of course. The work was no longer applicable to members so the study group didn't continue further. VT interest was to identify adhesives that could be used in ship outfit in steel ships, in particular, as an alternative to welding.

I came up with a different proposal for 'First Fixings', that of welding an array of special studs onto structural steel members to which could be bolted commercially available racking systems to which in turn equipments could be bolted. This eliminated the need to design special seats and their welding in place (hot work). But the saving didn't stop at reduced drawing and manufacturing. Should equipments be changed during build; often the case, a different supplier or design change from the original supplier, ne3w hot work would be needed to fit in the changed equipment, requiring removing insulation and anything that may catch fire during welding, and that could be on the deck above or the other side of the bulkhead. Not so with the new system; just change the bolt on racking to suit; job done. This system was patented in my name and I received £1000.

Before this could be patented and used; a lot of testing and development was needed to establish the limits of usefulness and approval from the MOD and Lloyds Register.

First a test rig was needed to bolt the racking onto studs, spot welded to an OBP standard steel section. This was set up in a Dartec Tensile Test Machine I had procured earlier for the new Mine-hunter build program to replace older equipment that had reached the end of it's useful life.

Using the UNISTRUT standard bolting and fixing arrangement we carried out pull off tests to discover areas of weakness that could easily be strengthened. UNISTRUT is a light racking system of U shape. The base is slotted for bolting. These tests showed the slots were prized over the standard bolt heads. A simple remedy was to use square load spreader plates under the bolt head; but a standard bolt could not be fitted with the spreader plate. This problem was overcome by using Cap Head UNBRAKO screws. The next weakness was found to be the standard Channel Nut; a threaded bridging piece; it bent before the channel section had reached full load capacity. There was an optional High Strength Channel Nut which we then used, and had a balanced design. These Hi Strength Channel Nuts were supplied as Specials with a Spiro Lock locking thread to prevent loosening under vibration.

The next step was to review the size and weight of Equipments throughout the ship to determine what was a sensible size and weight from the point of view of handling and acceptable shock withstand based on the static load test results.

The aim was a 25kg mass able to withstand a 25g shock. Torque tightening of the bolting had to be carefully specified to prevent over tightening and straining the joint and reducing the ductility of the fixing; an important feature to dampen out shock.

A test rig was made representing a typical section of bulkhead or deckhead, using a standard OBP section ,fitted with the Speccial Studs which also incorporated the Spiro Lock thread. A test box was also made representing a typical 25kg equipment. This test gear together with UNISTRUT parts were taken to a local Test & Vibration Centre at Titchfield for shock testing.

The test gear was set up in the shock test machine; making sure the bolt fixings were correctly torqued up. The set up was subjected to incremental shocks. After each shock the bolt torque settings were measured and re-tightened, so producing a curve of shock v torque loss, indicating the amount of shock deformation and damping in the joint. This series of tests was repeated with new UNISTRUT to obtain an average result.

Finally, the cabin mock-up was fitted with new UNISTRUT and incrementally tested without re-tightening the bolting until the capacity of the test machine was reached. Although loosened, the box was held firmly in place. Integrity of the joint had been proved.

The next phase was to develop outfit procedures. A typical ships compartment based on a corvette design was made in steel having all the structural reinforcement. It was fully studded out. UNISTRUT racking and fixings were provided and the Yard staff were invited to try out the new idea, and in so doing take part ownership.

The outfitted compartment was vibration tested to ensure no vibrational resonances occurred. The only 'fix' needed was to use thin distance pieces to offset the racking from direct contact with the base of the boxes. In this test, contact between the base and racking caused a noise rather like that of an electric train starting off!

Southampton University carried out Fatigue Tests on studded OBP sections on the insistence of Lloyds.

The system was now approved for application. A number of short seminars were run to get everyone up to speed in the firm who would use the 146rocedure.

It was decided the Triton Future Research Trimaran was to be the donor vessel. I wrote the Standard Procedures and work progressed. I also designed the Propeller Brackets, Superstructure and Funnel for this vessel.

This First Fixing Outfit System was subsequently used on all VT designed High Speed Attack Craft, the T45 Destroyer and Aircraft carriers, and most probably the Nuclear submarines. US Navy General Electric Nuclear Submarines used the concept. It was a step change in ships outfit.

The Modular Corvette Project was restarted. It had stopped when the other Engineers were transferred to work up a customer enquiry into a proposal and continue with that proposal till the end.

As part of the Modular Corvette team, I looked at Modular Cabin Design, and using a previous Corvette proposal incorporated the Modular Corvette Concept into a revised GA. It was far from a copy, but the proposal provided an excellent starting point. I also had to pull the Report together, incorporating the other Engineering Sections. My job was essentially that of Editor. A very interesting experience.

Type 45 Destroyer project definition work for the RN came along. I was put to work on Modular Cabin design. The idea behind a modular cabin is that a boxed in framework is designed, into which bunks, cupboards, maybe wet spaces (toilets and showers) etc. are fitted. When complete the modular cabins are delivered to the ship to be inserted into open ship hull modules, plumbed in and fixed in place before the ship module is taken to berth and welded into the ship.

That is the traditional cabin module build. But I couldn't help thinking there must be an easier way to build them. I began thinking of the car industry; they took fully fitted out doors to the car; not fit the door out in situ in the car. I thought, why not produce these cabins in segmental form to allow easy access during build. All that was needed would be to make the segments of L and T configuration, fully fitted out and taken to the ship and assembled into cabins there. An unexpected bonus was that this concept was not only lighter, it saved space as well.

It was too late for T45, but it was proposed for the Carriers. Whether that is how the cabin modules were done on these ships I never found out. Certainly, the weight and potential space saving (if you didn't use the saved space to make a larger cabin) on smaller ships would have been a priority on smaller ships.

I would like to have developed the segmental cabins further; there were interesting lighting proposals I was aware of that if a mock-up was built could have resulted, I felt in a step forward. I didn't press it, I was close to retirement and T45 took priority.

VT were involved together with BAE Systems on the design of an Anglo French Frigate. The superstructure was to be GRP, the hull steel. The French had patented a steel to GRP joint. The French didn't want to disclose the design data on their patented joint. It was my job to think of an alternative arrangement which VT could patent. This was done fairly quickly, but needed a lot of development. All VT did was to trial manufacturing the concept; which worked.

Faced with a possible alternative joint, the French agreed to disclose the design data of their joint. At this point no further work was undertaken on the VT Joint; or Geachean Joint as Strathclyde University called it! At that stage it was a production demonstrator, but more work was needed to develop and test the design.

My final R&D Project was on the Bonded Ship project. It was a European Project. The idea behind in was to bond honeycomb panels onto an al.alloy welded framed superstructure, divided into segments representing a superstructure on a large vessel.

The superstructure was designed to the Germanisher Lloyd Code.

Running into retirement I worked on the T45 Destroyer structure.

# 15. Making Your Own Luck.

Think about it; 'Making Your Own Luck'! It's true. Most of us have to find and make our own way.

To do this needs anticipation of what opportunities may be available; what qualifications and training will be needed to make that step up. And above all confidence not only in yourself, but that the job opportunities you want will come.

Nothing is for ever. The economy and business runs go in cycles. Position yourself to take that opportunity in the upturn when it occurs.

That's what I did, and it worked. I certainly didn't go to a fancy school; not even a good school; it can be done.

So, you aren't sure/don't know what you want to do! Well that's can be an advantage!

A tip; any good job will be interesting and challenging. What you need to know is the training on offer and where it may lead in that company, or elsewhere in that endeavour.

Keep an eye on the local papers, look at the business section. What firms have won new contracts or are advertising for staff? Apply and re apply.

Starting point as always is good training. Look for an Apprenticeship with a progressive company, look up firms in the telephone directory, research their business on the web so you know a bit about them.

Then go knocking on doors; see what's on offer. You would be unlikely to get an interview on that visit obviously, but use it to get an interview offer on another day. If unsuccessful there, keep going, don't get put off. Someone will offer you something.

A personal approach going door knocking; literally; is more effective than sending e-mails or letters; they are for the follow up after the interview; for confirmation/clarifications etc. should you be offered something.

Obviously the wage isn't negotiable, but aspects of training you may want to confirm is; such as the technical college course; whether you would have an introductory year, then depending on exam results be offered a place on the more academic courses, or maybe go straight onto a trade course. Or of course if you have good school exam results go into the academic course from the off.

You may well, and should have spoken of these things at interview, but if you were a bit tongue tied and didn't, or the offer in the letter wasn't clear on what course you would be studying, write in for clarification. And if this was discussed at interview, remind the interviewer of your understanding of what was said.

This brings me neatly to another point. Not only do some research about the company before interview, its products, does it have a design office etc, and therefore need Draughtsmen from time to time, but write out the questions you want to ask; for instance: transfer from craft to technician apprentice on certain attainment at technical college, what options would that then give you and so on. Express your ambition to say get into the Drawing Office. Make it realistic, a reasonable and credible step forward that the interviewer would be impressed with, and think to himself here's an ambitious person, worth taking on. That's how things work; that's what the interview is about; for the firm to assess you, and for you to assess the firm. Would it be a good fit? Both have to agree, so it's best foot forward. Your task at interview is to get an offer, then in your leisure decide if you want to accept or not. Don't go in ill prepared or half hearted.

Chance plays a big role. Keep plugging away, don't be put off.

In my day in the 1950's, apprenticeships abounded. They were proper 5 year craft apprenticeships then. Industry was expanding and so the opportunities for advancement were there.

These notes are slatted very much on my experience in Engineering, but the principles hold for whatever endeavour you choose. I've also addressed the school leaver more than the graduate, because I genuinely believe that today a University Education is no longer financially viable; possible gains in a university education over a technician apprenticeship with good technical qualifications are far outstripped by University fees and living expenses. Needs thinking about very hard indeed, unless you have sponsorship from a firm or relatively well off parents who can afford it, without placing themselves in jeopardy!

At the end of the day, the jobs aren't there needing all the graduates coming onto the employment market year on year. There never were; not even in the good years of strong economic growth where good jobs abounded. The graduate in many cases ends up disappointed, disillusioned and in serious debt.

What if with your best endeavours you can't find an apprenticeship that appeals. It's now a 4 year agreement, quite a chunk of life if your doing something you don't want to do. Often overlooked is the fact that it takes 2 or 3 years post apprenticeship to really become proficient in a technical trade.

Well you have to think 'out of the box'! If you can't get an apprenticeship and, say you want to become a draughtsman; depending on exam results you may need to retake a subject or two at Technical College to qualify for a BTECH Course in Mechanical Engineering. Limit Technical College to one day a week. BTECH's are one day a week for 3 years.

To pay for Technical College; for the remaining 4 days of the working week you can find a job; maybe Labouring with a small builder where you will learn basic brick laying and roofing, etc. depending on what jobs come up.

Do one evening class also learning CAD; in your first year. With a CAD qualification and first year of BTECH behind you, go looking for draughting work in small firms, say an Architects Office; and offer to work 3 months for free if they will train you, if you can afford it, or parents support you over that time. You would have to limit this to 4 days a week because of going to Tech. one day a week. If they offer a job at the end of the 3 months; Great, but don't be disappointed if they don't.

You may be interested in building and repairing computers. Same thing applies, offer to work free for 3 months. Study for appropriate qualifications at Tech one evening a week say. If they are busy you will probably get a job at the end of it. If not, keep plugging away.

What do you do for money? Maybe bar work at night; it shouldn't be for a long time. But that means living at home whilst getting on your feet.

Don't overlook what the armed forces apprenticeships can offer. You would get fit, have fine training, camaraderie, good qualifications and come out with a really good skills set.

Look into it; find out how long you have to sign on for to do what interests you. Go away, think about it and compare what is offered in civvy street.

# 16. The Best & Worst Firms

Marconi Space Systems was without a doubt the worst firm I ever worked for. In general it wasn't that bad, but I was in a toxic department. Force of circumstances made my position very difficult to impossible.

English Electric Aviation, F.G. Miles Engineering, Dowty Rotol, Rolls-Royce, New Zealand Electricity, Straughan & Henshaw and Vosper Thornycroft were all good firms and had much to commend them.

Technically it has to be the Rolls Royce. Stress Office.

For career development and wide interest; New Zealand Electricity.

For potential career development Straughan & Henshaw; provided they kept busy.

For spirit de corps and variety of interesting work; Vosper Thornycroft.

Dowty Rotol offered a wide range of interesting work. Ideal at that time of my development, I left too soon.

# 17. Reflections

In my younger years as an Apprentice, Aerospace was at a pinnacle; developing advanced military aircraft, airliners and missiles. Opportunities for the keen ambitious engineer abounded.

Company's were pleased to take you on and train you up. Technical College fee's were heavily subsidised, so they were affordable to those who wanted to get higher qualifications and progress.

I took full advantage of what was on offer. I got a push to get me started, when I transferred my apprenticeship to English electric aviation at Warton and into the Draughtsman's Course; I never looked back.

Came across envy and professional jealousy, but that's par for the course. Also met and worked with outstanding Engineers who taught me a lot.

The UK Aerospace Industry was brought to it's knees, when the newly elected Labour Government of the early 1960's elected on the 'White Heat of Technology' ticket, proceeded to cancel all the high tech aerospace projects; Supersonic Harrier, the 681 Battlefield STOL transport, TSR2 and I think Blue Streak. They tried to cancel Concorde, but the French said non! Instead the RAF bought Hercules Transports and Phantoms from the US. The Phantoms were re-engined with Rolls Royce Spey engines; Dowry Rotol designed a new front landing gear, providing greater angle of attack to improve take off from carriers.

Dowty Rotol lost out heavily, they had contracts to provide landing gear and systems to all the cancelled projects.

The Government then went into a huddle with the French, and came up with Anglo- French Variable Geometry (AFVG) bomber, replacing TSR2. When the French understood how the variable wing geometry worked, they pulled out and did it alone.

The Germans and Italians joined the UK project which became Tornado. In the process technology transfer and design responsibility was shared. It has been ever thus, resulting in the UK loosing civil airliner production of Airbus aircraft to Germany and France. Military production is limited to Warton where they assemble Typhoon and Hawk. I would be surprised if a viable design team is left. So much for our wonderful MP's!

www.ingramcontent.com/pod-product-compliance
Lightning Source LLC
Chambersburg PA
CBHW060856170526
45158CB00001B/374